Fast Techniques for Integrated Circuit Design

Do you want to deepen your understanding of complex systems and design integrated circuits more quickly? Learn how with this step-by-step guide that shows, from first principles, how to employ estimation techniques to analyze and solve complex problems in IC design using a simplified modeling approach. Applications are richly illustrated using real-world examples from across IC design, from simple circuit theory to electromagnetic effects and high-frequency design and systems such as data converters and phase-locked loops. Basic concepts such as inductance and capacitance are related to each other and to other RF phenomena inside a modern chip, enhancing understanding without the need for simulators. Use the easy-to-follow models presented to start designing your own products, from inductors and amplifiers to more complex systems.

Whether you are an early-career professional or researcher, graduate student, or established IC engineer looking to reduce your reliance on commercial software packages, this is essential reading.

Mikael Sahrling is Principal Engineer for ASIC Design at Tektronix Inc. He has over 20 years of experience in analog and mixed-signal circuit design.

"The estimation analysis techniques in this book open up a new and unique approach to gaining a deeper understanding of circuits, thus accelerating the optimization and design of a broad range of circuits, which is a critical skill in the fast paced IC design world where time to market is crucial to success."

Joel King, *Skyworks Solutions, Inc.*

"Developing engineering solutions benefits greatly from the proverbial back of the envelope analysis. This book does an excellent job of not only providing a great reference to a number of estimating techniques (limitations clearly identified) for a number of key topics. It also resurrects the concept of engineering estimation, to quickly evaluate ideas and drive to useful conclusions without losing context. This art form is dwindling as today's engineers continue to depend on (very capable) computer simulators, slowing the development of intuition and hence innovation."

Claudio Anzil, *Innophase Inc.*

Fast Techniques for Integrated Circuit Design

MIKAEL SAHRLING
Tektronix Inc.

CAMBRIDGE UNIVERSITY PRESS

CAMBRIDGE
UNIVERSITY PRESS

University Printing House, Cambridge CB2 8BS, United Kingdom

One Liberty Plaza, 20th Floor, New York, NY 10006, USA

477 Williamstown Road, Port Melbourne, VIC 3207, Australia

314–321, 3rd Floor, Plot 3, Splendor Forum, Jasola District Centre, New Delhi – 110025, India

79 Anson Road, #06-04/06, Singapore 079906

Cambridge University Press is part of the University of Cambridge.

It furthers the University's mission by disseminating knowledge in the pursuit of education, learning, and research at the highest international levels of excellence.

www.cambridge.org
Information on this title: www.cambridge.org/9781108498456
DOI: 10.1017/9781108682336

© Mikael Sahrling 2019

This publication is in copyright. Subject to statutory exception and to the provisions of relevant collective licensing agreements, no reproduction of any part may take place without the written permission of Cambridge University Press.

First published 2019

Printed in the United Kingdom by TJ International Ltd. Padstow Cornwall

A catalogue record for this publication is available from the British Library.

Library of Congress Cataloging-in-Publication Data
Names: Sahrling, Mikael, 1964– author.
Title: Fast techniques for integrated circuit design / Mikael Sahrling (Tektronix Inc.).
Description: Cambridge, United Kingdom ; New York, NY : Cambridge University Press, 2019. | Includes bibliographical references and index.
Identifiers: LCCN 2019009296 | ISBN 9781108498456 (hardback) | ISBN 1108498450 (hardback)
Subjects: LCSH: Integrated circuits–Design and construction.
Classification: LCC TK7874 .S187 2019 | DDC 621.3815–dc23
LC record available at https://lccn.loc.gov/2019009296

ISBN 978-1-108-49845-6 Hardback

Cambridge University Press has no responsibility for the persistence or accuracy of URLs for external or third-party internet websites referred to in this publication and does not guarantee that any content on such websites is, or will remain, accurate or appropriate.

For Nancy and Nicole

Contents

	Preface	*page* ix
1	**General Guidelines in Estimation Analysis in Integrated Circuits**	1
	1.1 Introduction	1
	1.2 Principles	1
	1.3 Integrated Circuit Applications	3
2	**Basic Amplifier Stages**	4
	2.1 Introduction	4
	2.2 Single Transistor Gain Stages	4
	2.3 Two Transistor Stages	18
	2.4 Summary	29
	2.5 Exercises	29
	2.6 References	30
3	**Higher Level Amplifier Stages**	31
	3.1 Introduction	31
	3.2 Five Transistor Amplifier	31
	3.3 Cascode Stage Amplification Using Active Feedback	33
	3.4 Comparator Circuit	34
	3.5 Cascaded Amplifier Stages	44
	3.6 Summary	48
	3.7 Exercises	48
	3.8 References	49
4	**Electromagnetism: Fundamentals**	50
	4.1 Introduction	50
	4.2 Maxwell's Equations	51
	4.3 Capacitance	66
	4.4 Inductance	72
	4.5 Various High Frequency Phenomena	85
	4.6 Summary	97
	4.7 Exercises	99
	4.8 References	100

5 Electromagnetism: Circuit Applications — 101
- 5.1 Introduction — 101
- 5.2 Connection to PCB Designs — 102
- 5.3 Recent Progress in the Literature on Signal Integrity On-Chip — 103
- 5.4 Transmission Line Theory — 105
- 5.5 S-Parameters — 109
- 5.6 Capacitors in Integrated Circuits — 119
- 5.7 Inductors in Integrated Circuits — 122
- 5.8 Design Examples — 140
- 5.9 Summary — 147
- 5.10 Exercises — 147
- 5.11 References — 148

6 Electromagnetic Field Simulators — 151
- 6.1 Introduction — 151
- 6.2 Basic Simulator Principles — 151
- 6.3 Long Wavelength Simulators — 152
- 6.4 Method of Moments — 159
- 6.5 Summary — 163
- 6.6 Exercises — 163
- 6.7 References — 163

7 System Aspects — 165
- 7.1 Introduction — 165
- 7.2 Jitter and Phase Noise — 166
- 7.3 Phase-Locked Loops — 171
- 7.4 Voltage Controlled Oscillators — 183
- 7.5 Analog-to-Digital Converters — 195
- 7.6 Summary — 231
- 7.7 Exercises — 231
- 7.8 References — 232

Appendix A Basic Transistor and Technology Model — 234
Appendix B Useful Mathematical Relationships — 239
Index — 242

Preface

This book is a result of many years in active design work in the semiconductor industry. I started my career as a theoretical physicist working on dense matter theory and electromagnetic fields in an astrophysical environment. After a few years my interest turned toward integrated circuit design, where there were also electromagnetic fields, and I have been working in this field ever since. It is a rich environment for the study of nature and mathematics and I am thrilled to be a part of it. As a theoretical physicist one always tries to get a handle on experiments and observations by doing simple mathematical modeling, and in my stint as a postdoc in the theoretical physics group at Caltech in the 1990s I was part of an Order of Magnitude Physics, 103c class that took this idea to town. The students were asked to estimate things such as the amount of rubber dumped into the air from cars on LA's freeways and how long a grass straw would grow in a week with a given precipitation and sunlight. The class was taught by Professor Peter Goldreich and Professor Sterl Phinney, and it opened my eyes to the power of estimation. In my career I have always tried to understand things by first estimating the impact of a certain effect and then verifying it. This analysis method has been a great help for me personally and the people I have been lucky enough to tutor. I have also encountered many other engineers and academic professionals who are very good at following these same principles. This book is an attempt to bring this way of thinking about design in general and circuit design in particular to a broader audience. I refer to the analysis method as estimation analysis, but many people use the term hand calculations, which I find to be rather misleading. Simply put, we consider complex problems in a way that do not require exact full solutions. The book will show that this approach can be taken for almost any problem, be it circuit analysis, high frequency phenomena, sampling concepts or jitter, to name a few. The scope of the book is from simple circuit theory, familiar to most engineers, to high frequency theory with a particular focus on integrated circuit applications, to systems such as data converters and phase-locked loops (PLLs). The applications are intentionally fairly broad, to illustrate the power of the techniques. What is different in this book compared with other similar ones is a strict physical approach where all situations are modeled carefully, often from first principles, followed by useful solutions and illustrative relationships after some algebra. Once such a model is established one can use it as a starting point for simulations where the simulator is used to fine-tune the design.

It is assumed that the reader is familiar with basic electromagnetism and circuit theory. There is no need to have any previous exposure to high level systems such as

PLLs and analog-to-digital converters (ADCs). Mathematical maturity corresponding to one year of college-level calculus and vector analysis is also assumed.

The book will start with a short chapter outlining the basic modeling concept, followed by two chapters describing basic circuit analysis where this modeling approach is used. These chapters should be familiar to most readers; the intention is to use the analysis technique in a familiar setting as an illustration. Then, in Chapters 4–6, the modeling concept is applied to high frequency situations, with a focus on integrated circuit applications. Here, we will also take the opportunity to define the concepts of capacitance and inductance in a way that shows their duality. Using the modeling technique, other interesting and much less discussed aspects of high frequency issues are further highlighted. The final chapter of the book describes higher system level applications where the same principles are applied. It covers PLLs and ADCs, their building blocks, and some of their properties. The hope is that this will help practicing electronics engineers reduce the need for simulators, and help them focus on the key problems faster. Each chapter also contains a set of exercises so that the reader can get more familiar with a concept and verify that the main points have been grasped.

The chapters are more or less independent of each other and for a one-semester class one can easily go through all the material. For students interested in electromagnetism and applications Chapters 4–6 should prove useful. For more system-level aspects of the estimation analysis technique Chapter 7 is a must. For a mild general introduction Chapters 2 and 3 on circuit analysis including nonlinear effects are of help.

A project like this cannot be completed without the help of many people. The patience and support of my family, Nancy and Nicole, have been unwavering. My manager, Pirooz Hojabri, and the technical staff at Tektronix have also been great champions of the project. In particular, Vincent Tso and Behdad Youssefi have read and commented on early drafts of the manuscript and Garen Hovakimyan and Patrik Satarzadeh have provided feedback on some of the mathematical derivations. I am very much indebted to them for their support and help. In addition, the helpful comments from the anonymous reviewers at Cambridge University Press helped elevate this book from a mere theoretical exercise to something much more useful for practicing engineers. Finally, I would like to thank the editorial staff at Cambridge University Press for their encouragement and support all through this project. Their sense of style and many helpful comments on the writing improved the book much beyond what I could have achieved alone.

1 General Guidelines in Estimation Analysis in Integrated Circuits

> Learning Objectives
> - Definition and overview of estimation analysis flow.

1.1 Introduction

This chapter provides a summary of steps needed to make useful mathematical models of physical systems. I refer to these steps as "estimation analysis" but in hallways of science and engineering schools or engineering offices they are often referred to as hand calculations or back-of-the-envelope calculations. I am not a big fan of these terms, as they convey a sense of sloppiness, which is far from accurate. This type of analysis is useful for building deeper understanding of integrated circuits and systems, but the methodology is very general and can be applied to most systems governed by some kind of mathematics. From deep space astrophysics to microscopic systems such as integrated circuits you will find broad applications of this kind of thinking. With this broad applicability it is no surprise that the principles we outline here are somewhat vague, but we will discuss enough examples in the rest of the book to enable the reader to develop a good sense of how to proceed in different situations. A mastery of these ideas will only come with experience. The process can be time-consuming initially because it involves digging into the core of the system under consideration. If the system is new to the user, the learning process can take even longer. But oftentimes, and with practice, the systems are similar enough to other systems the user has seen before that the process can be quite swift. We will start by outlining the principles and then discuss each of them in some depth. We will then refer to these steps in the following chapters, where many examples are provided.

After a model has been developed one can use it as a starting point for fine-tuning in a simulator or on the bench or whatever might be practical.

1.2 Principles

A beginner often tries to solve a problem with brute force, using three dimensions, full nonlinear equations, etc. The problem will then quickly become intractable with

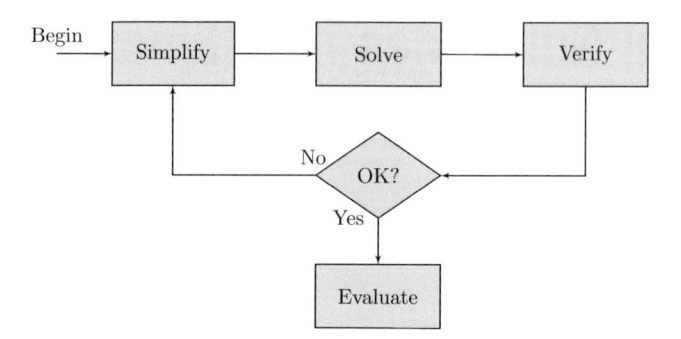

Figure 1.1 Estimation analysis work flow.

myriads of sums and complex expressions yielding little insight. With experience one learns that the core behavior is often much simpler to catch but it requires thinking the problem through before full calculations start. For the novice this can often be frustrating but with practice one learns to see the value of this approach.

In a typical modeling situation there are four steps to follow:

(1) Simplify – This is often the most difficult step because it attempts to get to the core of how the system works.
(2) Solve – If step 1 is executed properly this will be relatively easy.
(3) Verify – Here we verify the solution in step 2 is correct by for example checking extreme cases and/or comparing to simulations and/or exact calculations. If something is wrong, go back to step 1.
(4) Evaluate – In this section we analyze what the solution means.

We will discuss each of these in turn: see Figure 1.1 for a simple flow diagram.

Simplify

To properly simplify we need to be able to understand what we actually want to know. Is it a length scale, gain, bandwidth, linearity, etc.? How can we simplify the system so that this property will be highlighted? Can we make it two-dimensional, one-dimensional, cylindrical symmetry, spherical symmetry, planar symmetry? Can we linearize it? If we are interested in gain or noise, linearization is a great technique. Perhaps a harmonic tone is causing headache and perturbation calculation is in order? If we are looking at a clocked system, perhaps an ideal switch with infinite edge rate will suffice?

Solve

This often involves fairly simple algebra. In some cases there are some useful mathematical techniques available and in this book we will describe them as we move along and for more complex techniques we will refer to the literature.

Verify

This step is often neglected in practice but it is critical. If we have missed something essential in the model, the behavior of the model will simply be wrong and we will not have learned what we set out to learn. Oftentimes one can find similar calculations in the literature that can be used for verification. At other times one can simulate to confirm that the model fully captures the desired properties. This simulation is not to be used as a substitute for understanding, but to confirm the model assumptions and calculations are correct. A good technique is to take various parameters and go to the extremes to make sure the behavior is as expected. For example, if we are investigating the gain of an amplifier with a degeneration resistor, does the model go to the correct limit when this resistor is zero or infinite?

Evaluate

What does the solution mean? Is there something one can do with a certain variable combination that will result in something useful, like improved jitter or reduced power? If we need smaller inductance, will widening the metal width be efficient?

Most education comes with experience. The road is narrow and sometimes long but it is a great journey and the joy of understanding something on a deeper level cannot be overstated.

In this book we will illustrate estimation analysis methodology by going through many specific examples to showcase what can be done. My hope is that the reader will be inspired to go well beyond what is described here and invent great things. We will do simple circuit analysis that most readers will already be familiar with and then move on to complex examples such as the direct solution of Maxwell's equations and system analysis of analog-to-digital converters (ADCs) and phase-locked loops (PLLs). We will also discuss deeper physical concepts such as the nature of jitter and its relationship to phase noise.

1.3 Integrated Circuit Applications

Having established a simple model we can proceed to the design phase by applying the model to a specific design problem. We derive a set of parameters such as transistor sizes, interconnect width, etc. and use those as a starting point for the design in the simulator. The idea is to use the simulator to fine-tune the design. We should already know, within the accuracy of the simple model, what to expect from simulation. This kind of approach presents a huge shortcut in the design effort.

In this book we will build various models and use them in real-world design examples to establish a good starting point for fine-tuning in simulators.

2 Basic Amplifier Stages

> **Learning Objectives**
>
> - Applying estimation analysis to basic amplifier stages
>   - Linearization techniques – amplifier gain and impedance
>   - Perturbation analysis – weak nonlinear effects

2.1 Introduction

In this chapter we introduce basic transistor amplifier stages and use them as a starting point to describe the estimation analysis method. For more details on the transistors and the models used we refer to Appendix A.

For the reader familiar with the discussion in books such as [1–5] this chapter should not pose any difficulty.

Fundamentally, all we do in all these simplifications is to linearize the transistor or amplifier at its bias point and draw conclusions about fundamental properties such as gain and impedance. We ignore the body effect for clarity, by assuming that the body is always tied to source in all transistors.

In addition, we will also venture into the world of weak nonlinear effects and show how these gain stages can be analyzed with simple extensions to the standard linearization techniques, all in line with the estimation analysis method.

We start the chapter with a section on single transistor gain stages and continue with a few well-known two transistor stages. For brevity, we will focus on CMOS transistors, but other transistor types such as bipolar can easily be analyzed in the same way. We go through some design examples in detail in order to use the results in the later chapters.

2.2 Single Transistor Gain Stages

Single CMOS transistor gain stages are traditionally divided into three groups: common gate (CG), common drain (CD), and common source (CS) stages. The word common refers

2.2 Single Transistor Gain Stages

to the terminal that is common to both input and output signals, which can be either voltage or current. We will describe them one by one in this section. We will keep the discussion at a general level; the precise expression for the currents' dependence on terminal voltages does not matter. Only in the final expression, when we are after something specific, do we use specific current voltage relationships described in Appendix A.

CG Stage

The common gate (CG) stage is an amplifier where the gate node is tied to a fixed voltage, possibly with some impedance in series. The input signal enters through the source terminal and exits at the drain terminal. The signal is best described as a current.

Here we will solve for gain and input impedance.

Simplify We assume the transistor is in saturation so we will ignore the drain gate capacitance. We also assume the drain source impedance is sufficiently large so as not to affect the gain; and finally we assume the output load to be zero ohms. The transistor model in Figure 2.1 shows our assumptions. When comparing with the literature this could be seen as an over-simplification, but we are only interested in the dominant parameters that set the gain and input impedance so that the simplifications are an adequate approximation for an estimation analysis. To calculate the gain we will in addition linearize the transistor around its bias point.

We find for $Z_G = 0 \rightarrow v_g = 0$ and by applying Kirchoff's current law (KCL) at the output node

$$i_{out} = g_m \left(v_g - v_s\right) = -g_m v_s.$$

We also know

$$i_{in} = -v_s j\omega C + i_{out}.$$

Solve By solving for v_s we find

$$i_{out} = -\frac{g_m(i_{out} - i_{in})}{j\omega C},$$

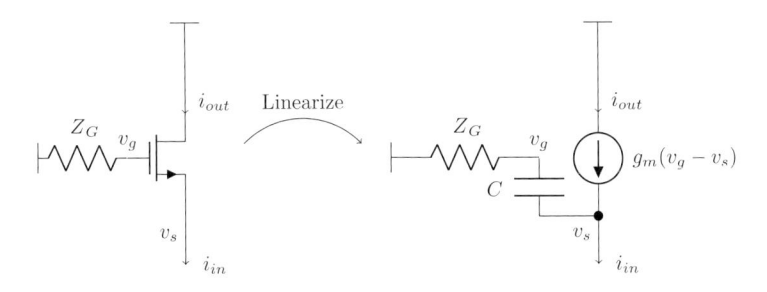

Figure 2.1 Common gate transistor stage with linearization.

or
$$\frac{i_{out}}{i_{in}} = \frac{g_m/j\omega C}{(1 + g_m/j\omega C)} = \frac{g_m}{(j\omega C + g_m)}.$$

We see for low frequencies that the input current goes straight through to the drain or output, but for higher frequencies the capacitor between the gate and the source will act as a short, effectively grounding the current and leaving nothing to the output. The transition point where $|j\omega C| = g_m$ is a rough estimate of the transition frequency or f_t. A more detailed model can be found in [3]. Here we get

$$f_t = \frac{g_m}{2\pi C}. \tag{2.1}$$

This is an important figure of merit for high speed designs and the expression (2.1) is a convenient rule of thumb.

What about the input impedance? We now have to rewrite i_{in} in terms of v_s:

$$i_{in} = -v_s j\omega C + i_{out} = -v_s j\omega C - g_m v_s = -(j\omega C + g_m) v_s.$$

We find
$$Z_{in} = -\frac{v_s}{i_{in}} = \frac{1}{j\omega C + g_m}.$$

Imagine now there is a gate impedance, Z_G. We find

$$v_g = -\frac{1/j\omega C}{1/j\omega C + Z_G} v_s + v_s = \frac{j\omega C \, Z_G}{1 + j\omega C \, Z_G} v_s,$$

or

$$i_{out} = g_m (v_g - v_s) = -g_m \left(\frac{1}{1 + j\omega C \, Z_G}\right) v_s.$$

To find the input impedance we need to rewrite i_{in} in terms of v_s.

$$i_{in} = (v_g - v_s) j\omega C + i_{out} = -\left(\frac{j\omega C + g_m}{1 + j\omega C \, Z_G}\right) v_s.$$

We the find after a simple rearrangement

$$Z_{in} = -\frac{v_s}{i_{in}} = \left(\frac{1 + j\omega C \, Z_G}{g_m + j\omega C}\right).$$

Verify As the reader no doubt recognizes, these calculations can be found in any standard electronics book, but here we have made some simplifications beyond that which is normally done. This is all in line with the estimation analysis idea. We are only seeking a model that is simple enough to capture the essence of what we want to know, in this case gain and input impedance. The calculations in this chapter are easy to verify in the literature. We will encounter more complex situations in Chapters 4–7.

Evaluate We have followed the estimation analysis method and we recognize the calculations from similar examples in the standard literature. To investigate the meaning of these expressions we need to go to various limits of key parameters. This is also often a way to sanity check the answer.

Let us look at the gain

$$A = \frac{I_{out}}{I_{in}} = \frac{g_m}{(j\omega C + g_m)}.$$

If $\omega \to 0$ we see the gain $A \to 1$. For high frequencies, $\omega \to \infty$ we see the gain $A \to 0$. Obviously, when $g_m \to 0$ the gain $A \to 0$.

Similarly, for the input impedance

$$Z_{in} = \left(\frac{1 + j\omega C \, Z_G}{g_m + j\omega C} \right).$$

We see an interesting relationship between the gate impedance and its reflection at the source. When $\frac{\omega}{2\pi} \ll f_t$ the gate impedance will be rotated by 90 degrees so a resistor in the gate will look like an inductor at the source, a capacitor will look like a resistor, and, most disturbingly, an inductor will look like a negative resistor, a kind of gain that can cause instabilities. For the limit $\omega \to 0$, the input impedance is simply $1/g_m$. In the other limit, $\omega \to \infty$ the input impedance is simply Z_G, which makes sense since in this case the gate capacitance shorts the $1/g_m$ from the transconductor.

CD Stage

For the common drain (CD) stage the input voltage goes to the gate of the transistor and the output is picked off of the source. It is often referred to as a source-follower circuit, or follower for short. The basic circuit configuration can be found in Figure 2.2.

We will solve for gain and input impedance.

Simplify First we will simplify the situation in a similar fashion to the CG stage.

Solve The output of the source will look like

$$v_{out} = (i_d + (v_{in} - v_{out})j\omega C)Z_L, \quad i_d = g_m (v_{in} - v_{out}).$$

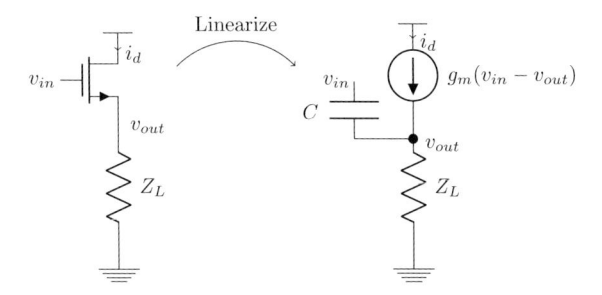

Figure 2.2 Common drain transistor stage with linearization.

Basic Amplifier Stages

After a simple rewrite

$$\frac{v_{out}}{v_{in}} = \frac{(g_m + j\omega C) Z_L}{(1 + (g_m + j\omega C)Z_L)}. \qquad (2.2)$$

The input impedance is now calculated by getting the input current

$$i_{in} = \frac{v_{in} - v_{out}}{1/j\omega C} = j\omega C \, v_{in} \frac{1}{(1 + (g_m + j\omega C)Z_L)},$$

and rearrange to find

$$Z_{in} = \frac{v_{in}}{i_{in}} = \frac{1}{j\omega C}(1 + (g_m + j\omega C)Z_L). \qquad (2.3)$$

Verify This is again a standard calculation in textbooks see for example [3].

Evaluate Let us look at the expression for gain, equation (2.2). When $\frac{\omega}{2\pi} \ll f_t$ we see

$$v_{out} = \frac{g_m Z_L}{(1 + g_m Z_L)} v_{in}.$$

For large load impedances, $g_m Z_L \gg 1$ $v_{out} \to v_{in}$. In the other extreme, $Z_L \to 0$ we get $v_{out} \to 0$, the output is simply shorted to ground.

As in the common gate stage we see the input impedance sees a 90 degree rotation of the impedance at the output, but this time it goes the other way: an inductor looks like a resistor, a capacitor looks like a negative resistor and a resistor looks like a capacitor. In fact for many input stages in narrow-band applications, like cellular phones, this property is a really nice way to create a low-noise input termination with the use of an inductor at the source of the input stage. The input stage will rotate this inductor to look like a real impedance with little noise(!). The remaining capacitor is often resonated out by a series inductor but that is a topic for another book.

CS Stage

The common source stage is perhaps a configuration that one often encounters early on in one's career. A common setup can be seen in Figure 2.3.

We will calculate gain and input impedance again.

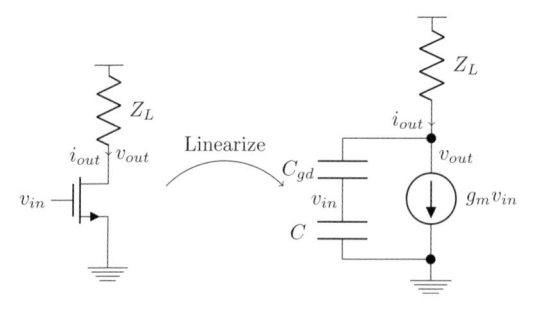

Figure 2.3 Common source transistor stage with output load.

2.2 Single Transistor Gain Stages

Simplify The output is a voltage when loaded with an impedance and the input is a voltage. We follow a similar linearization technique to that we had before, but this time we will include the gate drain capacitance, C_{gd}. We then have to solve KCL at the drain and source, we assume the gate driving impedance is zero.

Solve We have for the basic parameters

$$i_d = g_m(v_{in} - v_s) \quad v_s = 0 \quad i_s = i_d + j\omega C(v_{in} - v_s),$$

$$i_{out} = i_d + j\omega C_{gd}(v_{out} - v_{in}) \quad v_{out} = -i_{out}Z_L$$

We find

$$-\frac{v_{out}}{Z_L} = g_m v_{in} + j\omega C_{gd}(v_{out} - v_{in}) \rightarrow A_{gain} = \frac{v_{out}}{v_{in}} = \frac{j\omega C_{gd} - g_m}{1 + j\omega C_{gd} Z_L} Z_L.$$

The input impedance is

$$Z_{in} = \frac{v_{in}}{j\omega C_{gd}(v_{in} - v_{out}) + j\omega C v_{in}}$$

Upon substitution of v_{out} from the expression of gain above we can rewrite

$$Z_{in} = \frac{1}{j\omega C_{gd}\left(1 - Z_L(j\omega C_{gd} - g_m)/(1 + j\omega C_{gd} Z_L)\right) + j\omega C}$$

$$= \frac{(1 + j\omega C_{gd} Z_L)}{j\omega C_{gd}(1 + g_m Z_L) + j\omega C(1 + j\omega C_{gd} Z_L)}$$

Verify As before, this is a standard calculation in [2] but here we made even further simplifications to get an estimate of the gain and impedance.

Evaluate We see for low frequencies the gain, $A_{gain} = -g_m Z_L$, but there is a crossover frequency where the gain transitions at $\omega = g_m/C_{gd}$, in effect the major output current is supplied by the gate drain capacitance C_{gd} instead of the transistor gain. In the literature this is known as a right half plane zero. The input impedance is essentially a two-pole system due to the two capacitors. We see for low frequencies the total capacitance is the sum of C and $C_{gd}(1 + g_m Z_L)$, the gate drain capacitance has been amplified a factor $(1 + g_m Z_L)$. This effect is known as the Miller effect, the gain across a capacitor will amplify the capacitors value, increasing the effective load.

Nonlinear Extension

We can now employ the same technique to examine nonlinear extensions. In general one needs to employ Volterra series for electronics systems instead of the more commonly known Taylor series. This is due to the fact the systems we are considering have memory in that the output signal depends to some degree on what happened at earlier times in perhaps other parts of the circuit. The simple stages we will look at here have relatively high bandwidth in small geometry CMOS technologies, $f_t > 100$ GHz so we will assume a Taylor series expansion is appropriate. It often turns out to be quite useful, but care must be taken and the important verification step must be completed to make sure we do not fool ourselves and are better served with Volterra series.

For Taylor series we can write the output as a polynomial expansion of the input:

$$I_o = I_o^0 + I_o^1 V_o + I_o^2 V_o^2 + I_o^3 V_o^3 \ldots$$

Here

$$I_o^1 = \frac{dI}{dV}$$
$$I_o^2 = \frac{1}{2}\frac{d^2 I}{dV^2}$$
$$I_o^3 = \frac{1}{6}\frac{d^3 I}{dV^3}.$$

The coefficients can be calculated in several different ways: (1) One can sweep the DC bias point in a simulator and take the appropriate derivatives. (2) One can do a Fourier transform of the output when the input is a single tone sinewave, the linear and higher-order coefficients can be found from the harmonic powers. When relating the two methods keep in mind the mixing effect of the Taylor series when using sinusoids, $V_0 = A \sin \omega t$:

$$I_o = I_o^0 + I_o^1 V_o + I_o^2 V_o^2 = I_o^0 + I_o^1 A \sin \omega t + I_o^2 A^2 \sin^2 \omega t$$
$$= I_o^0 + I_o^1 A \sin \omega t + I_o^2 A^2 \frac{1 + \cos 2\omega t}{2} = I_o^0 + I_o^2 A^2 \frac{1}{2} + I_o^1 A \sin \omega t + I_o^2 A^2 \frac{\cos 2\omega t}{2}.$$

For example, the second-order term splits into a DC component and a second harmonic component. To find the size of the second harmonic term one needs to divide the second-order derivative of the transfer function by a factor of four. A factor of two comes from the Taylor expansion and another factor of two from the mixing action, where half the amplitude goes to DC and rest into the second harmonic. This is similar for higher orders but obviously more complex.

CD Stage

We will follow the CD stage discussion and make a first order correction to the gain calculation.

We start with

$$v_o = g_m (v_{in} - v_{out}) Z_L,$$

where we assume the frequencies of interest are far below f_t, and we are using Figure 2.2 for reference. In this section we limit ourselves to gain calculations.

2.2 Single Transistor Gain Stages

Simplify We will first simplify the discussion by assuming the drain source conductance is negligible. We use the following expression for

$$g_m = g_{m,0} + g'_m(v_{in} - v_{out}).$$

Finally, the load Z_L is for this discussion a real impedance.

Solve

$$v_{out} = \left(g_{m,0} + g'_m(v_{in} - v_{out})\right)(v_{in} - v_{out})Z_L = \left(g_{m,0}(v_{in} - v_{out})Z_L + g'_m(v_{in} - v_{out})^2 Z_L\right).$$

We now write

$$v_{out} = \alpha v_{in} + \beta v_{in}^2,$$

and put this into the expression

$$\alpha v_{in} + \beta v_{in}^2 = \left(g_{m,0}(v_{in} - \alpha v_{in} - \beta v_{in}^2)Z_L + g'_m(v_{in} - \alpha v_{in} - \beta v_{in}^2)^2 Z_L\right)$$
$$\approx \left(g_{m,0}((1-\alpha)v_{in} - \beta v_{in}^2)Z_L + g'_m(1-\alpha)^2 v_{in}^2 Z_L\right).$$

Now we identify terms of the same order on each side of the equal sign. We find

$$\alpha = g_{m,0} Z_L (1 - \alpha),$$
$$\beta = -g_{m,0} Z_L \beta + g'_m (1-\alpha)^2 Z_L.$$

We solve for

$$\alpha = \frac{g_{m,0} Z_L}{1 + g_{m,0} Z_L},$$
$$\beta = \frac{g'_m Z_L}{\left(1 + g_{m,0} Z_L\right)^3}.$$

And finally

$$v_{out} = \frac{g_{m,0} Z_L}{1 + g_{m,0} Z_L} v_{in} + \frac{g'_m Z_L}{\left(1 + g_{m,0} Z_L\right)^3} v_{in}^2.$$

Verify We verify this equation by simply simulating and varying the load, as in Figure 2.4. The basic transistor we are using is biased according to Tables A.2 and A.3 with a resistor at the load in addition to an ideal 1 mA bias current.

In effect the current/voltage bias of the transistor is unchanged as we vary the resistive load with its voltage termination so as to not draw additional bias current. We see from Figure 2.5 the agreement is reasonable for small values of the resistance Z_L but it gets increasingly worse for larger values. The reason for this is the nonlinear impact of the output resistance, r_o. In order to improve our predictive power we need to extend the analysis to take this effect into account also as we will show shortly.

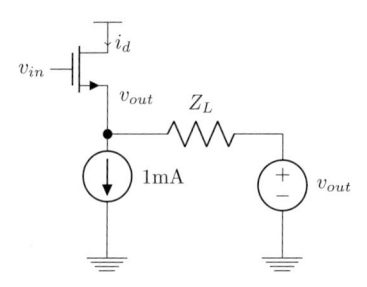

Figure 2.4 Simulation setup to verify gain versus load resistance with a fixed bias current.

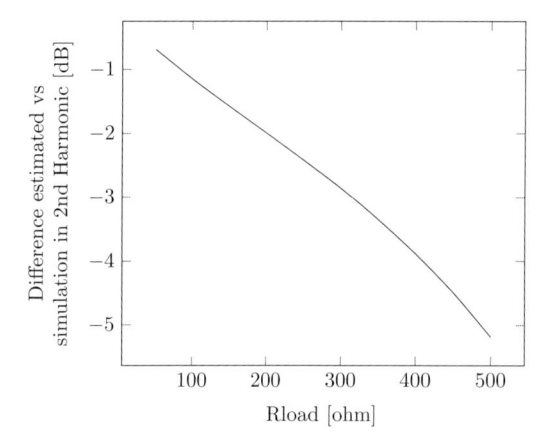

Figure 2.5 Comparison of estimation result to simulation.

Evaluate We see from the expression something that is somewhat intuitive. By increasing the load impedance we can reduce the effect of the second-order term.

Nonlinear Impact of Output Conductance

The output conductance is modeled as a resistor between source and drain of a transistor.

We will discuss how to calculate gain with this effect in mind.

Simplify We will extend the analysis to the case where the drain to source conductance is no longer negligible. We have in Figure 2.6 a simplified model where the drain voltage is an AC ground. The capacitance, C, is ignored

Solve Here we have the drain current dependent on two voltage differences, $v_{gs} = (v_g - v_s)$, $v_{ds} = (v_d - v_s)$. The Taylor expansion of the drain current needs to take both of these voltages into account including all second-order derivatives. We have

2.2 Single Transistor Gain Stages

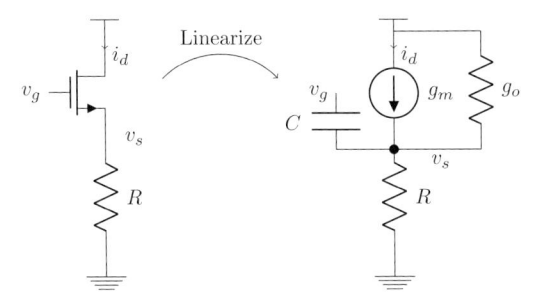

Figure 2.6 The output conductance model.

$$i_d = \frac{\partial i_d}{\partial v_{gs}} v_{gs} + \frac{\partial i_d}{\partial v_{ds}} v_{ds} + \frac{1}{2} \frac{\partial^2 i_d}{\partial v_{gs}^2} v_{gs}^2 + \frac{1}{2} \frac{\partial^2 i_d}{\partial v_{ds}^2} v_{ds}^2 + 2 \frac{1}{2} \frac{\partial^2 i_d}{\partial v_{gs} \partial v_{ds}} v_{gs} v_{ds}$$

$$= g_m v_{gs} + g_o v_{ds} + \frac{1}{2} \frac{\partial^2 i_d}{\partial v_{gs}^2} v_{gs}^2 + \frac{1}{2} \frac{\partial^2 i_d}{\partial v_{ds}^2} v_{ds}^2 + 2 \frac{1}{2} \frac{\partial^2 i_d}{\partial v_{gs} \partial v_{ds}} v_{gs} v_{ds}.$$

The derivatives need to be calculated at the bias point. We can ease the notation by using the following shorthand:

$$g'_m = \frac{1}{2} \frac{\partial^2 i_d}{\partial v_{gs}^2},$$

$$g'_o = \frac{1}{2} \frac{\partial^2 i_d}{\partial v_{ds}^2},$$

$$g'_{om} = \frac{1}{2} \frac{\partial^2 i_d}{\partial v_{gs} \partial v_{ds}}.$$

These three quantities can be either estimated or found by sweeping the DC bias point with a simulator. We have now

$$i_d = g_m v_{gs} + g_o v_{ds} + g'_m v_{gs}^2 + g'_o v_{ds}^2 + 2g'_{om} v_{gs} v_{ds}.$$

Using the expressions for v_{gs}, v_{ds} defined above and $i_d = v_s/R_l$ we find

$$\frac{v_s}{R_L} = g_m(v_g - v_s) + g_o(v_d - v_s) + g'_m(v_g - v_s)^2 + g'_o(-v_s)^2 + 2g'_{om}(v_g - v_s)(-v_s). \quad (2.4)$$

We now write

$$v_s = \alpha v_g + \beta v_g^2.$$

And keeping terms to second order:

$$\frac{\alpha v_g + \beta v_g^2}{R_L} = g_m\left(v_g - \alpha v_g - \beta v_g^2\right) + g_o\left(-\alpha v_g - \beta v_g^2\right) + g'_m\left(v_g - \alpha v_g\right)^2$$
$$+ g'_o(-\alpha v_g)^2 + 2g'_{om}\left(v_g - \alpha v_g - \beta v_g^2\right)(-\alpha v_g)$$
$$= g_m\left(v_g - \alpha v_g - \beta v_g^2\right) + g_o\left(-\alpha v_g - \beta v_g^2\right)$$
$$+ g'_m v_g^2 (1-\alpha)^2 + g'_o \alpha^2 v_g^2 - 2g'_{om} v_g^2 (1-\alpha)\alpha.$$

We now identify the like terms

$$a = g_m R_L(1-a) - g_o R_L a,$$
$$\beta = -g_m \beta R_L - g_o \beta R_L + g'_m R_L(1-a)^2 + g'_o R_L a^2 - 2g'_{om} R_L(1-a)a.$$

We find

$$a = \frac{g_m R_L}{1 + g_m R_L + g_o R_L},$$

$$\beta = \frac{g'_m R_L (1 + g_o R_L)^2 + g'_o g_m^2 R_L^2 - 2g'_{om} R_L (1 + g_o R_L) g_m R_L}{(1 + g_m R_L + g_o R_L)^3}.$$

Verify We simulate this situation and find the following comparison graph in Figure 2.7.

We have a much-improved comparison and can feel confident we have the correct model when including the nonlinearity of both the transconductance and output conductance.

Evaluate It is fairly intuitive to realize the output resistance r_o needs to be high to minimize this effect. In a normally biased CMOS transistor $g_m r_o \gg 1$, so the impact of a varying output resistance is less significant.

These examples have shown a common situation: one tries to estimate an effect which looks reasonable for certain parameter values but for others it falls short. The task is then to extend the model to cover the greater range by finding out the missing piece. In our case this was relatively simple but other situations can be much more complex. However, the same idea applies: Simplify → Solve → Verify → Evaluate.

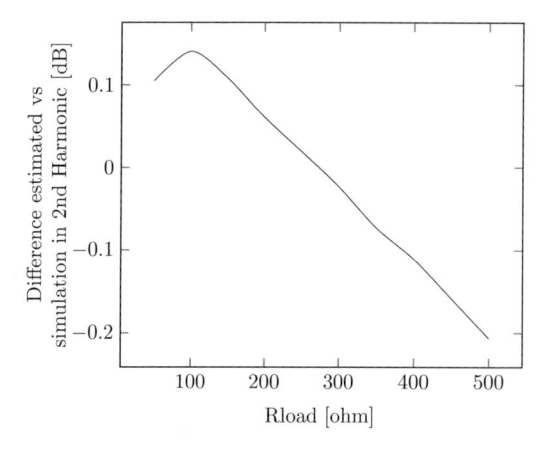

Figure 2.7 Comparison of estimation result vs simulation with the output conductance modeled.

CS Stage

The nonlinear extension to the CS stage can be analyzed along the same lines as the CD stage.

Simplify We make the same simplification as before with

$$g_m = g_{m,0} + g'_m v_{in}.$$

We also ignore all capacitors to make sure the Taylor expansion is valid.

Solve We write the drain current as

$$i_d = g_m v_{in} + g'_m v_{in}^2.$$

The output voltage is

$$v_{out} = -i_d R_L = -g_m v_{in} R_L - g'_m v_{in}^2 R_L.$$

Thanks to this simple model we now see that we have the nonlinear expression we were looking for.

Verify This calculation is a trivial extension of the simulation that gives us the expansion parameters in the first place.

Evaluate A low value of g'_m will result in a lower distortion. A more complex case is given by the case where there is a source resistance and we leave that exercise to the reader together with the output conductance variations with voltage swing.

Design Examples

We will use our knowledge of the transfer function and impedances of these amplifier stages to build and investigate some real-world examples.

Example 2.1 Estimate input impedance of a CD stage at 25 GHz

In this example we analyze the input impedance of a CD stage with a specific bias and a specific load. The load is another stage which looks capacitive and the interconnect has a, comparatively high, resistance. In effect the load looks like a resistor in series with a capacitor to ground according to Table 2.1. We will use this circuit in Chapter 7.

Solution
We assume the impedance of the current sink bias is large compared to the load impedance in Table 2.1 and we know from equation (2.3), assuming we are far below f_t,

Table 2.1 Specification table for CD stage

Specification	Value	comment
Type	Thin oxide NMOS	
Size (W/L/nf)	1 μm/27 nm/10	Biased in saturation according to Table A.3
Number of instantiations	2	
Load	$10 - j/(\omega\, 20\ \text{fF})$	Ohm

$$Z_{in} = \frac{1}{j\omega C}(1 + g_m Z_L) = \frac{1}{j\omega C}\left(1 + g_m\left(R_L + \frac{1}{j\omega C_L}\right)\right) = -\frac{1}{\omega^2 C C_L}g_m - j\frac{1}{\omega C}(1 + g_m R_L)$$
$$= -\frac{1}{\omega C_L}\frac{f_t}{f} - j\frac{1}{\omega C} - j\frac{f_t}{f}R_L.$$

where we have used the definition of f_t.

The operating frequency is about ten times less than f_t (see Appendix A), and we find then

$$Z_{in} = -\frac{1}{\omega C_L}10 - j\frac{1}{\omega C} - j10 R_L = -3180 - j678,$$

where we have used $C = 2 \cdot 2/3 \cdot 8$ fF ≈ 11 fF, which corresponds to a negative resistor with resistance of 3180 ohm and a capacitance of 9 fF at 25 GHz.

Example 2.2 CD stage with resistor capacitor ladder load
In this example we will calculate the transistor size needed to meet a certain output bandwidth. We assume the CD stage gate driving impedance is 0 ohms. The load is here a string of small resistors and at each resistor interconnect there is a capacitor to ground. For details see Table 2.2.

Solution
The load is a series of resistors where at each node connecting the resistors there is a capacitor to ground. The output impedance of the transistor itself is $1/g_m$. This impedance will for low frequencies drive the whole capacitive load. At higher frequencies the resistors in the series will eventually dominate the impedance and the gain response will be flat at the transistor source node. We have

$$\frac{1}{g_m}C = \frac{1}{2\pi f} < \frac{1}{2\pi f_{3\,\text{dB}}} \rightarrow g_m > 2\pi f_{3\,\text{dB}}\, C.$$

Plugging in the numbers we see

$$g_m > 2\pi \cdot 64 \cdot 8 \cdot 10^{-15} \cdot 25 \cdot 10^9 = 0.08\ \text{mho}.$$

We choose our unit transistor, Appendix A, Table A.3, with ten instances in parallel, this results in a current of 10 mA bias. What remains is to estimate the total impedance

2.2 Single Transistor Gain Stages

Table 2.2 Specification table for CD stage with ladder load

Specification	Value	Comment
BW 3 dB	>25 GHz	@ transistor source
Load impedance	R/C ladder with R = 625 mohm C = 12 fF, 64 units	
Gain	>−0.1 dB	
2nd harmonic distortion	<−40 dBc	200 mV input amplitude

needed to reach the specified gain. We know the gain from equation (2.2) and with the specification in linear units we find

$$\frac{g_m Z_L}{(1 + g_m Z_L)} > \text{Gain}_{spec}.$$

Where we have assumed $\omega \ll \omega_t$. Plugging in the numbers we find

$$\frac{0.1 Z_L}{(1 + 0.1 Z_L)} > 0.99 \rightarrow (1 - 0.99) 0.1 Z_L > 0.99 \rightarrow Z_L > 990 \text{ ohm}.$$

We will require the sink impedance to be larger than 1000 ohms to be on the safe side. The distortion will be small due to this large impedance of the current sink and from the second harmonic terms calculated in the section "Nonlinear Extension: CS Stage" we see the second harmonic term should be

$$H_2 \approx \frac{g'_m R_L (g_o R_L)^2 + g'_o g_m^2 R_L^2 - 2 g'' R_L g_o R_L g_m R_L}{(g_m R_L)^3} V_{in}^2 \sim \frac{1}{g_m R_L} V_{in}^2 = 10^{-3}$$

– in other words, less than 50 dBc, which easily meets our specifications.

Summary

We have looked at the common single transistor gain stages from an estimation analysis angle. We found by stripping away minor contributors to gain and impedance we end up with a simple model of the performance that can be used as a starting point for simulator fine-tuning. We also took a brief look at nonlinear extensions to the basic gain-transfer function and we found we can model the effect within a few fractions of a dB over a wide range of load resistance. The methodology can be summarized as one of simplify, solve, verify, evaluate and we showed a handful of examples where this procedure was followed. We also found a case where we needed to refine the model to capture a larger parameter range. The purpose was to show the reader the procedure in a familiar territory and we will venture out into less common areas in the following chapters. The biggest difference compared to earlier introductory classes is likely the much

Basic Amplifier Stages

simpler expressions. We have simply ignored parameters that have less influence on the properties we are studying.

2.3 Two Transistor Stages

We have looked at the classic single transistor gain stages. We will now expand to look at two transistor configurations. We start with the often used differential pair. We then investigate the classic current mirror and add a cascode to its output. As in the first section we will find here that these solutions are well known and we include them in the context of estimation analysis to increase the readers comfort level by showing how the process works in a familiar context.

The Differential Pair

The differential pair is a true work horse in the electronics industry: see Figure 2.8. There is hardly any integrated circuit that is manufactured that does not contain at least one such gain stage.

Simplify We will simplify the analysis by assuming all signals are antisymmetric around the center point. This point will become what is known as a virtual ground. Its voltage is constant and is in the AC sense a ground: see Figure 2.9. With this simplification we have

Solve We immediately recognize this as a CS stage we just investigated in the previous section. The only difference is that the input voltage is half of the input voltage to the full two transistor circuit. The output differential trans conductance is now trivially:

$$i_{out} = \frac{v_{in}}{2} g_m - \left(-\frac{v_{in}}{2} g_m\right) = v_{in} \, g_m.$$

which is again identical to our CS-stage analysis, ignoring the input capacitance and source impedance.

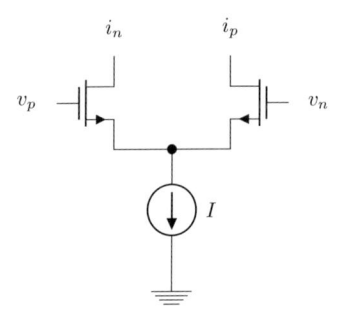

Figure 2.8 Differential pair without degeneration.

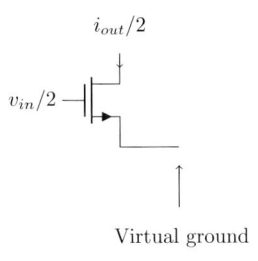

Figure 2.9 Symmetrized differential pair.

Verify A classic calculation that can be found in for example [2].

Evaluate We have seen that by utilizing symmetry one can greatly simplify the circuit analysis.

Nonlinear Extension

For a differential pair there is an interesting subtlety concerning the third harmonic. Consider this model of a differential pair where each leg has a drain current

$$i_d = g_m(v_g - v_s) + g'_m(v_g - v_s)^2. \tag{2.5}$$

If we approximate the current sink at the source node with an ideal current sink and ignore any degeneration resistor, the sum of the signal currents in both legs needs to be zero.

$$i_d^p + i_d^n = g_m(v_g^p - v_s) + g'_m(v_g^p - v_s)^2 + g_m(v_g^n - v_s) + g'_m(v_g^n - v_s)^2$$

$$= \left\{\text{using } v_g^n = -v_g^p = -v_g\right\}$$

$$g_m(v_g - v_s) + g'_m(v_g - v_s)^2 + g_m(-v_g - v_s) + g'_m(-v_g - v_s)^2$$

$$\approx -2g_m v_s + 2g'_m v_g^2 = 0.$$

Where we have assumed $v_s^2 \ll v_g^2$. We then have

$$v_s = \frac{g'_m}{g_m} v_g^2.$$

For $v_g \sim \sin \omega t$, we find the source voltage contains the second harmonic of the input signal, v_g. This is also easy to convince oneself of with the intuitive idea the source voltage 'sees' a pull-up/down from each leg and it occurs at twice the frequency. We have established the common source point sees a second harmonic. This is a well-known result for differential pair operations. Let us know look specifically at the second-order term in the expression for i_d equation (2.5), while assuming $v_g = A \sin \omega t$

Basic Amplifier Stages

$$g'_m(v_g - v_s)^2 \approx g'_m(v_g^2 - 2v_g v_s) = g'_m\left(v_g^2 - 2\frac{g'_m}{g_m}v_g^2\right).$$

The last term contains the third harmonic! It is being mixed up by the second-order expansion coefficient. This third harmonic can be significantly larger than the one resulting directly from the third order expansion coefficient. It is an interesting artifact of the differential pair operation. Be aware the inherent limitation in this analysis in that we assume there are no circuit elements with memory, like caps and inductors present so a Taylor expansion applies.

Current Mirror

Imagine now another extraordinarily common topology, the current mirror in Figure 2.10.

The gain is straight forward and we leave it to the reader to analyze. We will look at the noise transfer of this circuitry. We assume the reader is familiar with the concept of noise and has studied various modeling methods in other courses. For details on our noise modeling see Appendix A.

Simplify We simplify the circuitry by ignoring any capacitance and assuming there are two noise sources from drain to source for each transistor, as in Figure 2.11.

Solve The noise transfer for transistor M_1 is now simply

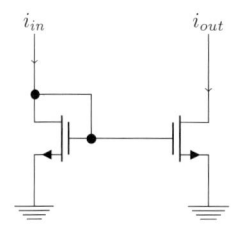

Figure 2.10 Current mirror.

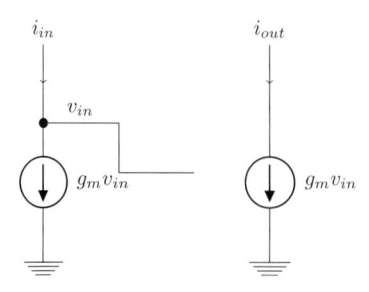

Figure 2.11 Linearized current mirror.

$$i_{n,out,1} = \frac{i_{n,1}}{g_{m,1}} g_{m,2}.$$

From the other transistor we now find trivially

$$i_{n,out,2} = i_{n,2}.$$

The two noise sources are uncorrelated so their power will add. We find

$$i_{n,out}^2 = i_{n,out,1}^2 + i_{n,out,2}^2.$$

Using the common assumption that $i_n = 4kT\gamma g_m$ we have

$$i_{n,out}^2 = 4kT\gamma \left(g_{m,1} + g_{m,2} \right).$$

Later we will look at the output impedance for this configuration and how it can be improved. We will therefore quickly realize the output impedance is

$$Z_o = r_o. \tag{2.6}$$

Verify This calculation can be found in [1–3].

Evaluate The key lesson here we will use in later calculations is that the noise *powers add if they are uncorrelated*. Since a common assumption for transistor noise is that its power is $\sim g_m$ we have the resulting noise is proportional to the sum of the transistors g_m for a simple current mirror configuration.

Simple Cascode Transistor

The output impedance we just studied is often inadequate. A common remedy is to add another CG stage at the output of the current mirror transistor.

Let us look at Figure 2.12 and investigate the output impedance

Simplify We do the same simplifications as earlier and linearize around the bias point: Figure 2.13.

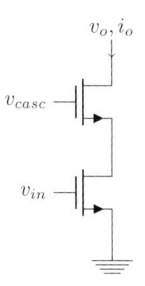

Figure 2.12 Simple cascode transistor stage.

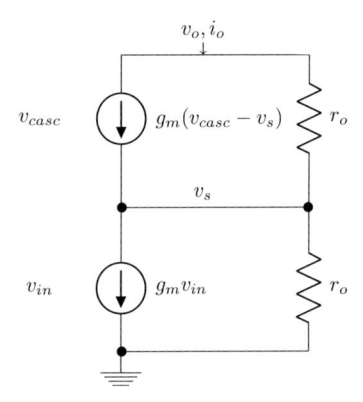

Figure 2.13 Linearized cascode transistor stage.

We also assume the transistors are such the output resistance r_o is the same for both transistors.

Solve We know that without the cascode transistor the output impedance is given by (2.6) where r_o is the output resistance of a single transistor. With the cascode we find by injecting a current at the output terminal and setting up KCL

$$\frac{v_o - v_s}{r_o} + g_m(-v_s) = i_o,$$

$$v_s = i_o r_o.$$

Combining we find

$$\frac{v_o - i_o r_o}{r_o} + g_m(-i_o r_o) = i_o,$$

$$i_o(2 + g_m r_o) = \frac{v_o}{r_o}$$

Finally

$$Z_o = (2 + g_m r_o) r_o. \tag{2.7}$$

Verify This calculation can be found in for example [2].

Evaluate The output impedance has been amplified by $(2 + g_m r_o)$. In the next chapter we will see another way of boosting the output impedance even further. We will see an example later on in this chapter where the cascode is made up of a thin oxide transistor which can be made much smaller in area than the bottom transistor for a given current, resulting in a reduced capacitive load. This type of cascoding is a very common arrangement to improve isolation and impedance.

CMOS Inverter

The CMOS inverter is another classic, see Figure 2.14.

This circuit is usually studied with propagation delay in mind [4, 5]. We will study its input impedance over frequency and its gain around the trip point in this section. In the next section we will see how a cross-coupled inverter operates around the trip point.

Simplify We simplify by looking at the gain around the trip point where the output is equal to the input and we will use the following simplified model shown in Figure 2.15.

The bias of the transistors is such that each is in saturation and the channel charge does not contribute to the gate-drain capacitance. However, we include such a capacitor here since the fringing capacitance can be a significant portion of the overall capacitance for small geometry CMOS technologies. We can now calculate gain and input impedance:

Solve Let us first look at the DC gain, by ignoring all capacitors. We find

$$i_l = \left(-g_m^p - g_m^n\right)v_{in}.$$

And

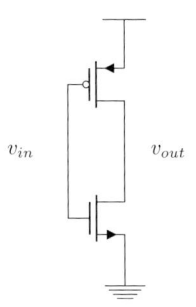

Figure 2.14 A CMOS inverter stage.

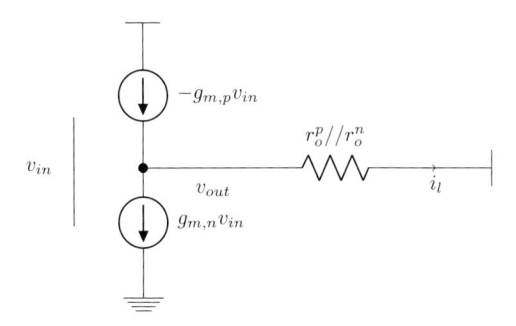

Figure 2.15 A simplified CMOS inverter around the trip point.

Basic Amplifier Stages

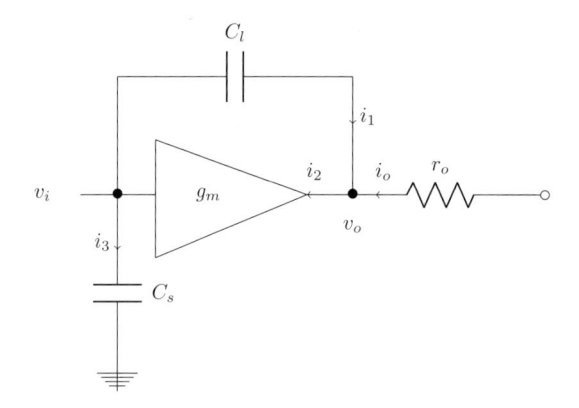

Figure 2.16 Simple model of an inverter where the average transconductance is used.

$$v_{out} = i_l(r_o^p//r_o^n) = (-g_m^p - g_m^n)v_{in}(r_o^p//r_o^n) = -g_m r_o v_{in}. \qquad (2.8)$$

Where we have assumed g_m is the average trans conductance of the NMOS and PMOS transistors and $r_o = 2(r_o^p//r_o^p)$, see [5].

The input impedance at DC is practically infinite (not counting the small leakage current due to tunneling). For the input impedance at some higher frequency we can ignore the shunting capacitor to ground in the first analysis and add it in later. We find now we can use the simple model in Figure 2.16 where we use the average transconductance of the two transistors to calculate the gain and impedance. The capacitance for the inverter amplifier is a complicated structure involving even nonlinear entities. Here we will simplify this by assuming the capacitors are constants.

For gain we have from the figure

$$v_o = -i_o r_o, \qquad (2.9)$$

$$i_1 + i_o = i_2, \qquad (2.10)$$

$$i_1 = C_l \frac{d(v_i - v_o)}{dt}, \qquad (2.11)$$

$$i_2 = v_i g_m, \qquad (2.12)$$

$$i_3 = C_s \frac{dv_i}{dt}. \qquad (2.13)$$

We here use (2.10)–(2.12) to get

$$i_1 + i_o = C_l \frac{d(v_i - v_o)}{dt} - \frac{v_o}{r_o} = v_i g_m,$$

and after rearranging:

$$-\frac{dv_o}{dt} C_l - \frac{v_o}{r_o} = v_i g_m - \frac{dv_i}{dt} C_l. \qquad (2.14)$$

2.3 Two Transistor Stages

We have this in the time domain at this point since we will use that description later. In frequency space we have

$$-j\omega C_l v_o - \frac{v_o}{r_o} = v_i g_m - j\omega v_i C_l,$$

giving

$$\frac{v_o}{v_i} = \frac{-g_m + j\omega C_l}{j\omega C_l + 1/r_o}. \tag{2.15}$$

The input impedance can be calculated from

$$i_1 = j\omega \left(1 - \frac{-g_m + j\omega C_l}{j\omega C_l + 1/r_o}\right) v_i C_l = j\omega \left(\frac{g_m + 1/r_o}{j\omega C_l + 1/r_o}\right) v_i C_l,$$
$$i_3 = j\omega C_s v_i$$

We find

$$\frac{v_i}{i_1 + i_3} = \frac{(j\omega C_l r_o + 1)}{j\omega C_l(g_m r_o + 1) + j\omega C_s(j\omega C_l r_o + 1)}. \tag{2.16}$$

Verify This trip point transfer function, (2.15), can be found in for example [5], the rest of the calculations and the derivation of the differential equation can verified either with a simulator or symbolic manipulation software. The dominant time constant is given by $r_o \cdot C_l$, in line with for example [4]. We also find from (2.15) at low frequencies we recover (2.8). For low frequencies the input impedance (2.16) goes to infinity as we expect, for high frequencies the impedance goes to zero, the shunt capacitance C_s simply shorts out the amplifier. All this is in line with what one should expect from the model.

Evaluate The bandwidth of the inverter is in this approximation simply the RC time constant of the output load resistance and the cross-capacitance in this model. The input impedance approaches zero for high frequencies.

Cross-Coupled CMOS Inverter

We have taken a quick look at the basic CMOS inverter stage. We will now use what we learned to study a somewhat more complicated system; the cross-coupled inverter. We include here both the input/output shunting capacitor and the input capacitor to ground.

Simplify We have the following simple description shown in Figure 2.17:
This simplification starts to break down when the inverters become unbalanced but as an initial study of such circuits behavior it will prove quite insightful and we will use it in Chapter 3.

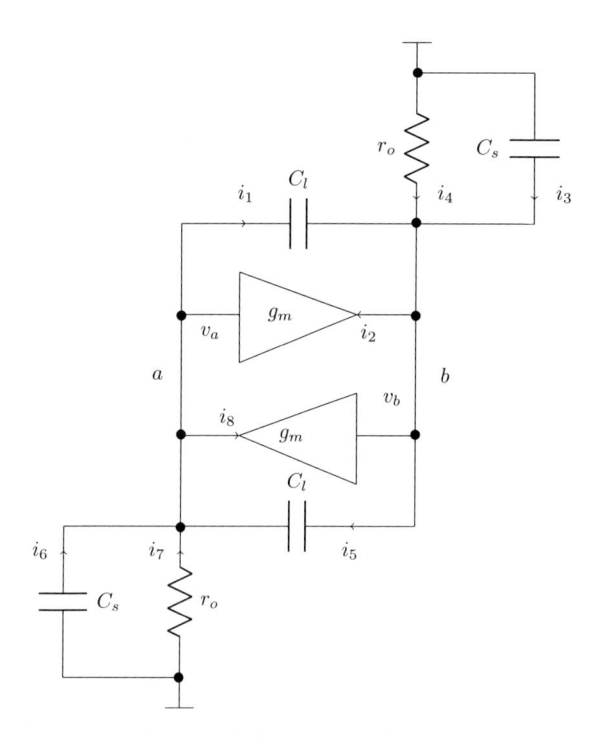

Figure 2.17 Cross-coupled inverter and its simplified model with an input/output shunting capacitor and a capacitor to ground.

Solve We will first setup the basic loop equations with the labels in the figure.

$$i_1 + i_8 = i_6 + i_7 + i_5, \tag{2.17}$$

$$i_5 + i_2 = i_4 + i_3 + i_1, \tag{2.18}$$

$$i_1 = \frac{d(v_a - v_b)}{dt} C_l = -i_5, \tag{2.19}$$

$$i_2 = v_a g_m, \tag{2.20}$$

$$i_8 = v_b g_m, \tag{2.21}$$

$$v_a = -i_7 r_o, \tag{2.22}$$

$$v_b = -i_4 r_o. \tag{2.23}$$

$$i_3 = -\frac{dv_b}{dt} C_s \tag{2.24}$$

$$i_6 = -\frac{dv_a}{dt} C_s \tag{2.25}$$

We have 10 unknowns and 10 equations so we should be able to make some progress here. First we see there is an antisymmetry in the problem in that

$$v_a = -v_b, \quad i_4 = -i_7, \quad i_3 = -i_6, \quad i_2 = -i_8$$

We find now

$$i_1 - i_2 = -i_3 - i_4 - i_1, \tag{2.26}$$

$$v_a = i_4 r_o, \tag{2.27}$$

$$i_1 = 2\frac{dv_a}{dt} C_l, \tag{2.28}$$

$$i_2 = v_a g_m, \tag{2.29}$$

$$i_3 = \frac{dv_a}{dt} C_s \tag{2.30}$$

we see by combining (2.26)–(2.30)

$$2i_1 = 4\frac{dv_a}{dt} C_l = i_2 - i_3 - i_4 = v_a g_m - \frac{dv_a}{dt} C_s - \frac{v_a}{r_o}$$

A simple rewrite now shows

$$\frac{dv_a}{dt}(4C_l + C_s) = v_a \left(g_m - \frac{1}{r_o}\right),$$

which has the solution

$$v_a = K e^{t/\tau}, \tag{2.31}$$

where

$$\tau = \frac{4\,C_l + C_s}{g_m - 1/r_o}. \tag{2.32}$$

Notice the factor of 4. It results from the simple loop and is akin to the Miller effect. The shunt capacitance, C_l corresponds to the gate drain capacitance, C_{gd} and C_s is the gate souce capacitance C_{gs}. In saturation we know from elementary text books $C_{gs} \approx 2/3\, C_{ox}$. If we look at the fringe gate-drain capacitance in the appendix we see it is about 1/2 of the gate-source capacitance. The total capacitance in the numerator of equation (2.32) is then

$$4\,C_l + C_s = 4\frac{1}{3} C_{ox} + \frac{2}{3} C_{ox} = 2 C_{ox} \tag{2.33}$$

Finally the timescale for our process and using the transistor in Appendix A is

$$\tau = \frac{2 C_{ox}}{g_m - 1/r_o}. \tag{2.34}$$

Verify This can be easily verified in a simulator.

Basic Amplifier Stages

Table 2.3 Specification table for current mirror with cascode

Specification	Value	Comment
Voltage compliance	>500 mV	
Output resistance	>1000 ohm	
DC current	10 mA	

Table 2.4 Starting point sizes for current sink

Device	Size	Comment
M_1, thin oxide	W/L = 200 µm/30 n	
M_2, thick oxide	W/L = 200 µm/200 n	
Output resistance	1.25 kohm	$V_{out} = 500$ mV

Evaluate For small geometry CMOS the fringe capacitance gate-drain cannot be ignored. In fact its effect is also amplified by a Miller like phenomenon in cross-coupled pairs. Of interest here is the fact the timescale is not set by $r_o \cdot C$ like we had earlier, the gain of one of the stages $r_o g_m$ speeds up the loop and we are left with a much shorter timescale.

Design Examples

Example 2.3 **Current mirror with cascode and large output impedance**
In design Example 2.2 we defined the needed output impedance of the current sink for the follower (CD) stage we designed. Let us here design a current sink that can sink the needed current and at the same time provide the required output impedance: for specification see Table 2.3.

Solution
We know from this technology that a minimum size device has $g_m r_o \approx 10$, see Appendix A. With a simple cascode the mirror transistor can then have an output impedance that is 10 times larger than the mirror output impedance by itself. We see from the appendix that a thick oxide transistor with $l = 200$ nm might suffice: it has an output resistance of about 2.5 kohm. It needs at least 300 mV to be in saturation, Figure A.4. The cascode transistor is then left with 200 mV drain source voltage and from Figure A.2 it seems this is adequate if the gate-source voltage is ~470 mV. The total sink current is 10 mA and this means we can bias a transistor at $V_g \sim 770$ mV and $m = 20$ instances to get the required current draw for an output resistance of 125 ohm.

We find as a starting point the parameters in Table 2.4:

The simulator output is shown in Table 2.5. The output resistance is 1.1 kohm at 500 mV which is slightly above our specification. The cascode transistor does not have a $g_m r_o$ that is quite 10, hence the difference from our estimate.

Table 2.5 Final sizing table after simulation optimization

Device	Size	Comment
M_1, thin oxide	W/L = 200 μm/30 n	
M_2, thick oxide	W/L = 200 μm/200 n	
Output resistance	1.1 kohm	$V_{out} = 500$ mV

With this current sink design we have together with Example 2.2, a full CD stage we can use in Chapter 7.

2.4 Summary

We have looked at the common transistor gain stages from an estimation analysis angle. We found by stripping away minor contributors to gain and impedance we end up with a simple model of the performance that can be used as a starting point for simulator fine tuning. We also took a brief look at nonlinear extensions to the basic gain-transfer function and we found we can model the effect within a few fractions of a dB over a wide range of load resistance for a particular configuration. The idea can be summarized as one of simplify, solve, verify, evaluate and we showed a handful of examples where this procedure was followed. The purpose was to show the reader the procedure in a familiar territory and we will venture out into less common areas in the following chapters.

2.5 Exercises

1. Investigate common mode properties, gain and output impedance, of the differential pair. Verify!
2. Calculate third-order correction to the CD stage. Verify with the following

$$v_o = \frac{-g_m R_l}{1+g_m R} v_{in} + \frac{-g'_m R_l}{(1+g_m R)^3} v_{in}^2 + \frac{2g_m'^2 R_l R - g''_m R_l(1+g_m R)}{(1+g_m R)^5} v_{in}^3.$$

3. Calculate the gain and input impedance of the CS stage where the transistor has a source degeneration impedance. Verify!
4. a. Derive the second-order correction to the CS stage with a degeneration resistor at the source. *Hint*, compare with CD stage.
   b. Include the nonlinearity of r_o.
5. In the cross-coupled inverter model calculate the differential admittance by injecting a current between the nodes *a*, *b* in Figure 2.17 and calculate the

resulting voltage across the pair. This expression will be useful in the later chapters. $\left[\text{Result } \frac{i}{v} = -g_m + \frac{1}{r_o} + j\omega \left(C_{sh} + 4C_{cr}\right)\right]$.

2.6 References

[1] D. A. Johns and K. Martin, *Analog Integrated Circuit Design*, Hoboken, NJ: Wiley, 1996.
[2] R. Gray, J. Hurst, S. Lewis, and R. Meyer, *Analysis and Design of Analog Integrated Circuits*, 5th edn., Hoboken, NJ: Wiley, 2009.
[3] B. Razavi, *Design of Analog CMOS Integrated Circuits*, 2nd edn., New York: McGraw-Hill, 2016.
[4] R. J. Baker, H. W. Li, and D. E. Boyce, *CMOS Circuit Design, Layout and Simulation*, 3rd edn., New York: IEEE Press, 2010.
[5] A. Sedra and K. C. Smith, *Microelectronic Circuits*, 7th edn., Oxford, UK: Oxford University Press, 2014.

3 Higher Level Amplifier Stages

> **Learning Objectives**
>
> - Applying estimation analysis to more complex amplifiers
>   - Gain calculations – cascaded amplifier stages
>   - Noise transfer
>   - Circuit description in the time domain – comparator

3.1 Introduction

We will discuss somewhat more complex amplifier configurations in this chapter. Following the treatment in Chapter 2 we will here analyze the amplifiers following the estimation analysis method. The calculations will be very similar to the standard literature in, for example, [1–3] and they serve more to showcase the methodology in a familiar setting than to demonstrate any new insights.

We start with the well-known five transistor amplifier, a classic interview question, and continue with cascode stage amplification using feedback. This is followed by a comparator discussion where we put the emphasis on simple timescales and noise analysis. After that, we investigate cascaded stages and the implication in terms of gain, noise, and linearity. Finally we show a couple of design examples that will be used in later chapters where several full-blown design examples will be built and the building blocks we have developed with simple models will be of help in designing the final circuits.

3.2 Five Transistor Amplifier

The classic five transistor amplifier is shown in Figure 3.1. It is used all over the circuit world in all manners of applications. We will here focus on noise transfer but we start with a quick description of its operation.

Simplify First we will simplify by assuming the various transistors have the same combination of g_m, r_o. Then we assume the current bias tail transistor has zero output

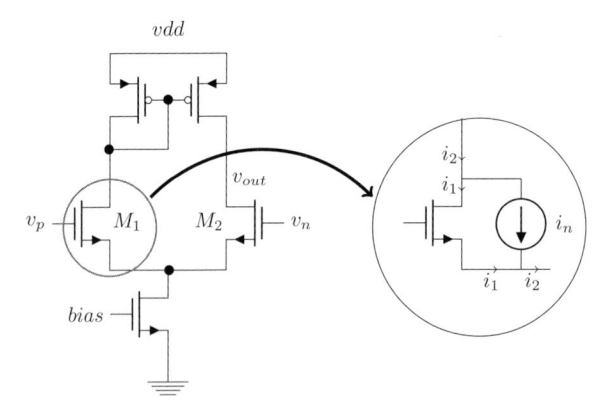

Figure 3.1 Five transistor operational amplifier.

conductance. The amplifier output resistance is modelled as a resistor $r_o = r_{o,p}//r_{o,n}$ between the output node and supply.

Solve The basic operation of the amplifier is as follows: The transconductance gain of the differential pair is causing a current to go into the load. The load is a simple current mirror and the current from transistor M_1 is appearing at the output out-of-phase with the corresponding current generated by transistor M_2. The output voltage swing is finally given by the output current times the load, r_o. We find using $v_{in} = v_p - v_n$

$$v_{out} = \frac{v_{in}}{2} g_m 2 r_o = v_{in} g_m r_o$$

and the gain

$$A = \frac{v_{out}}{v_{in}} = g_m r_o.$$

Verify This is a classic calculation that can be found in [3] for example. Let us now look at noise transfer. This is a common interview question.

Simplify
- The noise from all the transistors is uncorrelated so the noise powers will add
- The noise voltage is small so a linearized version of the circuit suffices to capture the noise transfer function.
- The NMOS transistors have the same transconductance, $g_{m,n}$ and PMOS has $g_{m,p}$

Solve We will solve this by calculating the noise transfer from each of the noise sources and add their power at the output.

Calculating KCL at both the source and the drain shows half the noise current goes through the opposite transistors source node into the PMOS mirror, the other half goes through the same transistor and closes on itself, see the zoomed-in portion of Figure 3.1. The loop formed by the noise current that goes through the opposite transistor causes the current to be mirrored by the PMOS load and the two half currents add up at the output node. We find

$$i_{n,1}^{out} = \frac{i_{n,1}}{2}\frac{g_{m,p}}{g_{m,p}} + \frac{i_{n,1}}{2} = i_{n,1} = \sqrt{4kT\gamma g_{m,n}},$$

$$i_{n,2}^{out} = \frac{i_{n,2}}{2}\frac{g_{m,p}}{g_{m,p}} + \frac{i_{n,2}}{2} = i_{n,2} = \sqrt{4kT\gamma g_{m,n}} = i_{n,1}.$$

The PMOS current mirroring is much simpler, we have from the section "Current Mirror" in Chapter 2 $\left(i_p^{out}\right)^2 = 2 \cdot 4kT\gamma g_{m,p}$.

Finally, the current bias transistors current splits into two where one half goes through the PMOS mirrors the other directly to the output load resistor. Keeping track of the sign of the current we find

$$i_{n,bias}^{out} = \frac{i_{n,bias}}{2} - \frac{i_{n,bias}}{2} = 0.$$

The total output noise voltage now becomes

$$v_{n,o}^2 = \left(2 \cdot 4kT\gamma g_{m,n} + 2 \cdot 4kT\gamma g_{m,p}\right)r_o^2 = 8kT\gamma \left(g_{m,n} + g_{m,p}\right)r_o^2.$$

And input-referred,

$$v_{n,i}^2 = \frac{v_{n,o}^2}{g_{m,n}^2 r_o^2} = \frac{8kT\gamma}{g_{m,n}}\left(1 + \frac{g_{m,p}}{g_{m,n}}\right)$$

where we have assumed the correction factor γ, is the same for both NMOS/PMOS.

Verify See for instance [3, 4] where this problem is discussed in some detail.

Evaluate The final expression here looks deceptively simple. The key realization here is the noise currents for the differential pair transistors actually splits evenly and gets transferred to the output through two different paths.

3.3 Cascode Stage Amplification Using Active Feedback

In Chapter 2 we saw how a cascode transistor improved the output impedance of a current mirror. We can go one step further and we will now show an even higher impedance can be achieved with active feedback. Let us look at Figure 3.2. The amplifier senses the cascode source voltage and amplifies it to drive the gate of the cascode transistor. We will next analyze this within the estimation analysis framework we have been using.

Simplify Let us simplify by assuming the gain has infinite bandwidth. We leave to the reader to solve the finite bandwidth case.

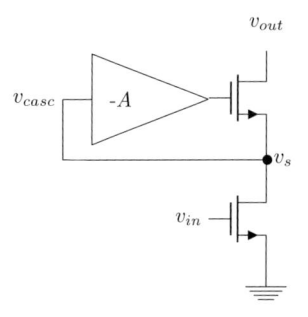

Figure 3.2 Cascode stage with an amplifier in feedback.

Solve Similarly to the previous example we have

$$\frac{v_o - v_s}{r_o} + g_m(v_g - v_s) = \frac{v_o - v_s}{r_o} + g_m(-Av_s - v_s) = \frac{v_o - v_s}{r_o} - g_m v_s(A + 1) = i_o.$$

The only difference to before is the gain $A + 1$ in front of the transconductance. The result is

$$Z_o = (2 + (A + 1)g_m r_o)r_o.$$

Verify This calculation can be found in [2] for example, where the active cascode case is investigated in detail.

Evaluate As with the previous example we see a great additional impedance boost can be achieved with a cascode transistor, this time using an additional amplifier in a loop configuration.

3.4 Comparator Circuit

A comparator circuit is shown in Figure 3.3. This so-called strong-arm comparator is a work horse in modern integrated circuit data converters. It has gained this popularity due to its low-power, a few mW is not unusual, high gain and speed. We will analyze the circuit in a few steps where we will employ the estimation analysis to each. It is expected that with such a popular circuit topology there is plenty of analysis in the literature, see [5, 6], to just mention a few and we will intentionally be somewhat brief in our discussion here.

Comparator Analysis

The analysis can be naturally simplified by dividing it into three phases the circuit goes through as a function of time:

3.4 Comparator Circuit

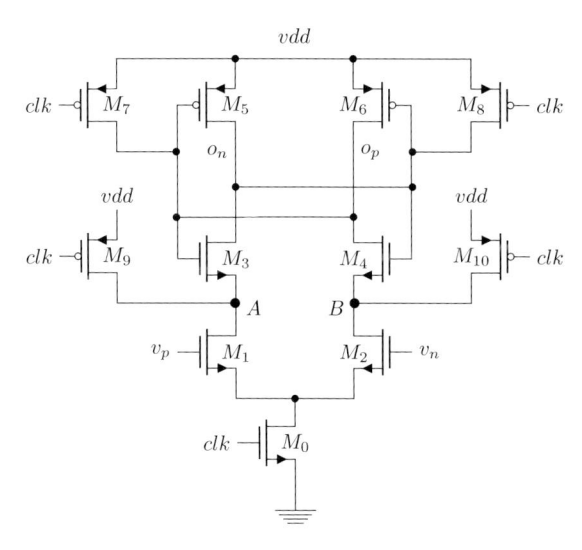

Figure 3.3 Strong-arm comparator circuit.

- Reset phase. Here transistors $M_7 - M_{10}$, turns on pulling the nodes, o_p, o_n, A, B to *vdd*. Also the switch, M_0, at the tail turns off the differential pair so the rest of the nodes can be pulled high to *vdd*.
- Initialization phase. The reset voltage goes high enabling the differential pair to be active and releasing the rest of nodes o_p, o_n, A and B. The nodes A, B start to get pulled down by the differential pair until first the NMOS transistors, M_3, M_4 turns on and then the output nodes o_p, o_n, start to get pulled down until the PMOS transistors M_5, M_6 gate voltage goes below its threshold. The nodes A, B continues down to ground, in effect shorting out the input pair.
- Regeneration phase. Here the input pair is disconnected from the circuit operations and the output stage cross-coupled inverters start to decide if the input is high or low.

We will mostly ignore the reset phase here. This is more due to space limitation than to any prejudice against it. We will discuss the last two stages briefly following the steps in the estimation analysis.

Initialization Phase

In the initialization phase the nodes start from their reset voltages, *vdd* and moves depending on the input voltages more or less quickly to the point where the top PMOS transistors turn on. Here we will first show a possible way to simplify this stage and capture some of its characteristics. We then solve this simplified model and compare to simulations.

Simplify Let us simplify the operation of this stage by looking at Figure 3.4. We have removed the tail switch and just look at one side of the circuit. We will estimate the timescales to discharge the capacitances at A, o.

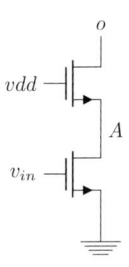

Figure 3.4 Initialization phase of the strong-arm comparator.

Solve The timescale needed to discharge a capacitance can be found from the governing differential equation, and we will show a simple example here. From basic text books we know that for a capacitor with capacitance C its charge

$$Q = CU,$$

where U is the voltage across the capacitor. Taking the derivative with respect to time gives

$$\frac{dQ}{dt} = I(t) = C\frac{dU}{dt}(t).$$

where $I(t)$ is the current through the capacitor. We can then estimate the timescale, τ, by approximating $dU/dt \approx \Delta U/\tau$ and find

$$\tau = C\frac{\Delta U}{I}.$$

For our circuit we see that node A gets discharged due to the current going through transistor M_1. Its capacitance, C_A is set by the combined junction capacitance of M_1 and M_3. When the voltage at A reaches $vdd - V_t$, transistor M_3 turns on. We find

$$\tau_{i,1} = \frac{V_t C_A}{I_b} \qquad (3.1)$$

where V_t is the threshold voltage of transistor M_3. The current now continues to discharge node A until it reaches ground and will also discharge the output node, O, until node O goes down enough so that the PMOS transistor turns on. Assuming node A is at ground this results in a timescale for the output node to be discharged of

$$\tau_{i,2} = \frac{V_t C_o}{I_b} \qquad (3.2)$$

where we assume the threshold voltages for transistors M_3 and M_5 are the same. In this chapter we will only look at various limits and we will assume the relevant timescale for the initialization phase is either (3.1) or (3.2) depending on the situation. See Figure 3.5a and b for a simulation of the initialization phase, where the input nodes are at the same voltage.

3.4 Comparator Circuit

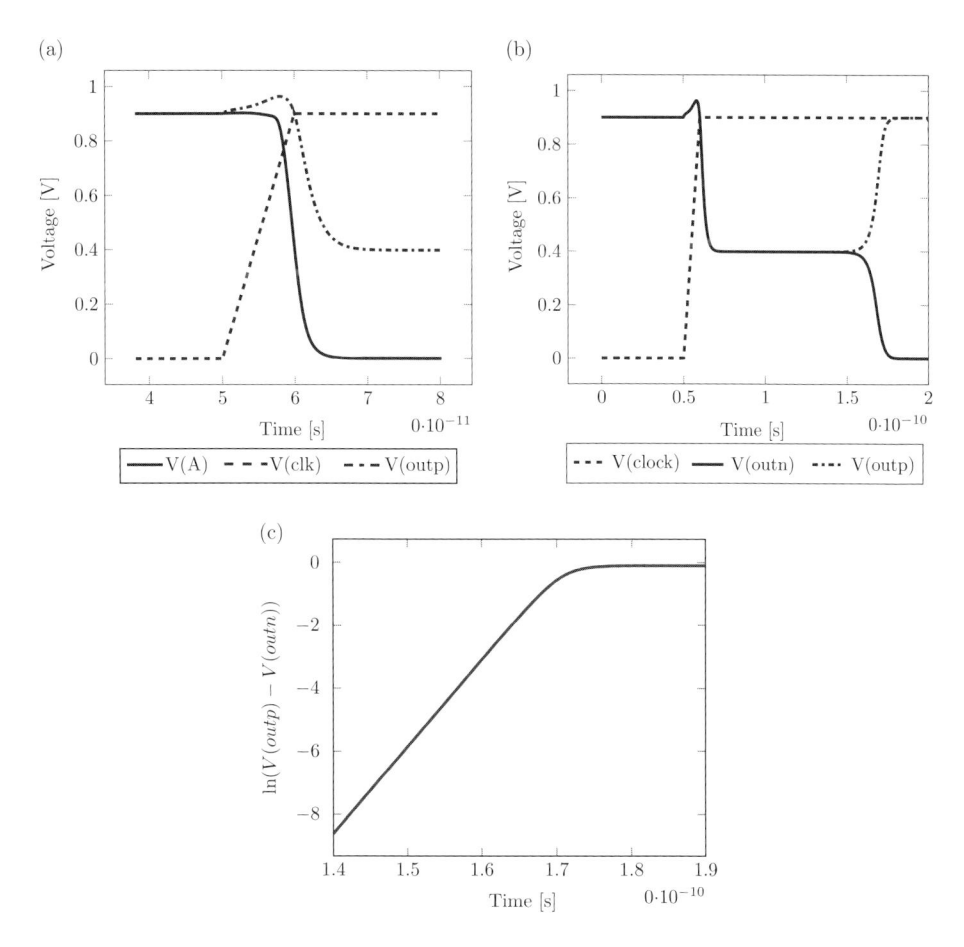

Figure 3.5 Simulation of comparator decision sequence where in (a) and (b) various node voltage are displayed. Figure (c) shows the logarithm of the output differential voltage to indicate the regeneration time scale.

Regeneration Phase
At this stage the transistors start to look like cross-coupled inverters and we will use the results from the discussion in the section "Cross-Coupled CMOS Inverter" in Chapter 2. We know from this section that the time evolution of the system varies like

$$v_o \sim e^{t/\tau_r}$$

where
$$\tau_r = \frac{C_o}{(g_{m,n} + g_{m,p})/2}. \tag{3.3}$$

The operation can be easily verified as in Figure 3.5c, where the regeneration cycle can be clearly seen.

Verify This case is studied in more detail in [5].

Evaluate We see from the estimates of timescales that we need to have a large input stage to generate sufficient current and a low capacitive load in order to reduce the time needed to make a decision. With a capacitive load, C_{load}, at the output and the regeneration timescale τ_r from (3.3) shows it is directly dependent on the output capacitance. We need to make sure we have enough transconductance, g_m in the cross-coupled pair to drive it. With this output load we get using $C_o = C_{load} + C_{self}$, where the C_{self} can be found from equation (2.33).

$$\tau_r = \frac{2C_o}{(g_{m,n} + g_{m,p})} = \frac{2C_{load} + 2C_{self}}{(g_{m,n} + g_{m,p})}.$$

The self-capacitance $C_{self} = 2C_{ox} \sim WL$, where we have used the channel width, W, and length, L of the cross-coupled transistors. $C_{load} \sim (W \cdot L)|_{load}$ is the load capacitance scaling with its transistor size. We know from elementary textbooks on transistor operation that $g_m \sim W/L$. We then see

$$\tau_r \sim \frac{(WL)|_{load} + WL}{W/L} = \frac{L}{W}(WL)|_{load} + L^2.$$

Without the output load we should use minimum length devices for optimal regeneration speed. To minimize the overall timescale the capacitance needs to be minimized leading to minimum width W of the cross-coupled device. With a specific load we see that a minimum channel length, L, is beneficial. The width W should be as large possible to overcome the load capacitance, but will be limited by the capacitance it offers the input stage.

Metastability

The positive feedback nature of the output results in an exponential increase in the output voltage with time

$$v_o = v_{start} e^{\frac{(t-t_0)}{\tau_r}} = v_{start} e^{\frac{t_d}{\tau_r}}$$

where we have defined the decision time $t_d = t - t_0$. Imagine now that $v_{start} = 0$. This would result in $v_o = 0$ indefinitely. It is called a metastable condition, we will investigate the impact of this phenomenon next with estimation analysis.

Simplify We will ignore any noise, thermal or other. We will assume there is a certain voltage v_x that if the crosscoupled pair output exceeds this voltage the following circuitry can operate properly. We will also annotate that given a certain decision time t_d and regeneration time τ_r, the input voltage needed to reach v_x in time t_d is v_c. We have

$$v_x = v_c e^{\frac{t_d}{\tau_r}}.$$

We will denote by v_{FS} the full-scale voltage at the input (see Figure 3.6) and finally v_{LSB} is the least-significant-bit voltage. If the input is within the gray region the output is uncertain.

We will now solve for the bit error rate.

3.4 Comparator Circuit

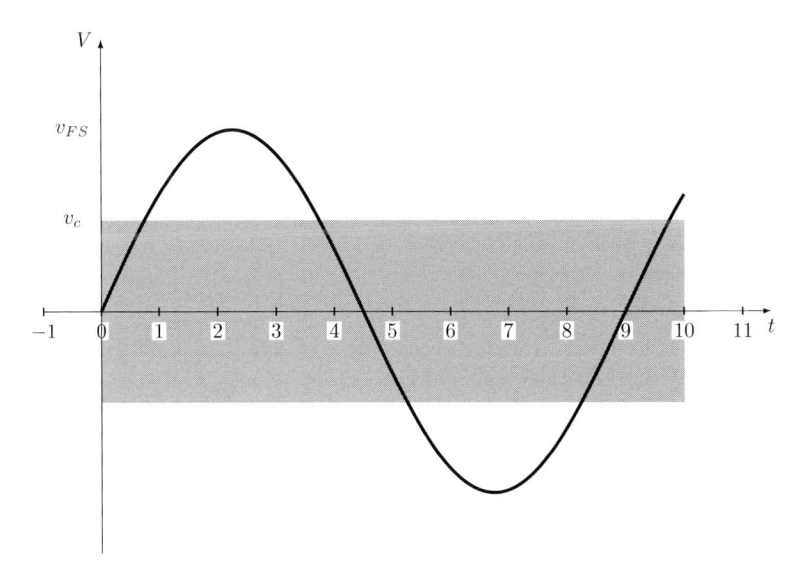

Figure 3.6 Input signal with uncertain decision region indicated in gray.

Solve Imagine we adjust t_d or τ_r in such a way as to reduce v_c by a factor of two. The chance of making wrong decision is now also reduced a factor of two, in that they gray region in Figure 3.6 is half in size. We can then motivate the following formula for the error probability:

$$P(E) \sim \frac{v_c}{v_{FS}} = K_m \frac{v_c}{v_{FS}}.$$

We need to define the constant K_m and we observe that if $v_c = 0.5$ *LSB* we will make a wrong decision half the time, or in other words an error rate of $\sim 10^0$. The constant K_m needs to be $v_{FS}/(0.5 v_{LSB})$. After some rewrite we have for

$$P(E) = \frac{v_c}{0.5 v_{LSB}} = \frac{v_x}{v_{FS}/2^{N+1}} e^{-\frac{t_d}{\tau_r}}, \tag{3.4}$$

where N is the number of bits in the system.

Verify This calculation can be found in most standard ADC texts like [8, 9].

Evaluate We can conclude the longer we allow the comparator to operate the smaller the bit error rate will be. This is hardly surprising but the exponential nature of the process is an important feature. The pay-off is much higher if one can adjust the exponent than the factor in front.

Input Pair Size
The input pair should be as large as possible while meeting the maximum load requirement at the input. At some point the input pairs own capacitance will dominate its load and further increase in size gives no benefit.

Reset Switch Size

Finally we need to set the reset switches. Fortunately, their operation is relatively straightforward. When the switch is ON it discharges the capacitor in question and the timescale is simply set by the slew rate

$$\tau_{reset} = \frac{vdd}{I_{switch}} C_l \sim \frac{L_{switch}\, vdd}{W_{switch}} C_l$$

where C_l is an appropriate load. As with the cross-coupled inverter this transistor also needs to have as short a channel length as possible and as long width as possible. The width will be limited by the load it represents to both the output of the decision circuit and the input transconductor of the comparator.

No Output Load

Perhaps a more interesting discussion is the question of how fast one can make this topology for a given technology. Let us look at the various timescales again, but this time we assume there is no load at the output. The input stage still needs to be as large as possible and we will assume its load is dominated by its own capacitance. We have then

$$\tau_i = C_A \frac{V_t}{I_b}$$

$$\tau_r = \frac{2C_{self}}{(g_{m,n} + g_{m,p})}.$$

For the basic transistor model and parameter relationships, please see Appendix A. We use

$$I_b = K \frac{W}{L}(V_G - V_t)^2$$

where V_G, V_t are assumed given. Also

$$C_{self} = 2C_{ox} = 2K_1 W \cdot L,$$

We use for the capacitance at node A the junction capacitance at its drain

$$C_A = K_2 W$$

$$g_{m,n} \approx g_{m,p} = 2K \frac{W}{L}\left(\frac{vdd}{2} - V_t\right).$$

The timescales can now be estimated as

$$\tau_i = \frac{V_t K_2 W}{K \dfrac{W}{L}(V_G - V_t)^2} = \frac{V_t K_2 L}{K(V_G - V_t)^2}$$

$$\tau_r = \frac{2K_1 W_c L^2}{2K W_c \left(\dfrac{vdd}{2} - V_t\right)} = \frac{K_1 L^2}{K\left(\dfrac{vdd}{2} - V_t\right)}.$$

Plugging in the process number for this particular technology from Appendix A, we see

3.4 Comparator Circuit

$$K = 0.7 \cdot 10^{-4} \left[\frac{A}{V^2}\right] \quad K_1 = 30 \cdot 10^{-3} \left[\frac{F}{m^2}\right] \quad K_2 = 2 \cdot 10^{-10} \left[\frac{F}{m}\right]$$
$$V_t = 350 \text{ m[V]} \quad V_G = 500 \text{ m[V]}$$
$$L = 30 \text{ [nm]}.$$

We find
$$\tau_i = 1.5 \cdot 10^4 \left(\frac{0.4 \cdot 2 \cdot 10^{-10} \cdot 3 \cdot 10^{-8}}{10^{-2}}\right) = 3.6 \text{ ps}$$
$$\tau_r = 1.5 \cdot 10^4 \left(\frac{30 \cdot 10^{-3} \cdot 9 \cdot 10^{-16}}{0.1}\right) = 4 \text{ ps}.$$

Comparing to the simulator we are in the same ball park $\tau_i \sim 3$ ps, and $\tau_r = 3.6$ ps, which is close to our estimate.

This does not take reset time into account which is roughly

$$\frac{L_{switch} \, vdd}{KW_{switch}(V_g - V_t)^2} \; C_l \approx \frac{L_{switch} \, vdd}{KW_{switch}(V_g - V_t)^2} \; K_2 W = 10^4 \cdot 1.5 \frac{3 \cdot 10^{-8}}{3 \cdot 10^{-6} \cdot 0.25} 2 \cdot 10^{-10} \cdot 10 \cdot 10^{-6}$$

$$= 10^4 \cdot 1.5 \cdot 10^{-2} 4 \cdot 10^{-16} \cdot 20 \sim 1.2 \text{ [ps]}.$$

To these estimates we need to add finite clock rise/fall times.

To see where we are qualitywise we can estimate the bit error rate. Let us assume that all timescales are zero except the regeneration timescale τ_r. Let us also assume the sampling rate is 25 GSps or sampling period is 40 ps. Assuming further we have a 6-bit system with a full scale voltage $v_{FS} = 800$ mV and a decision voltage $v_x = 200$ mV, we find from (3.4)

$$P(E) = \frac{0.2}{0.8/128} e^{-40/4} \approx 10^{-3}.$$

In reality we need to allow some time for the rest of the steps. A typical error rate in communcation systems might be 10^{-6} so we are hard pressed to achieve 25 GSps for these transistors in this technology with a reasonable bit error rate. We can improve the situation by allowing for longer sampling period, using a lower V_t device or some other such scheme. For instance with a 12.5 GSps sampling frequency we find a minimum bit error rate of

$$P(E) = \frac{0.2}{0.8/128} e^{-80/4} \approx 10^{-7}.$$

This looks like a better option but recall this would mean we need a duty cycle that is far from 50 percent and the rest of the timescales need to be small.

Comparator Noise Analysis

We will now analyze the noise transfer of the strong-arm comparator following the steps we just used to estimate the fundamental timescales. We will focus specifically on noise transfer and ultimately signal to noise ratio (SNR). We will assume the input signal is sampled already and is constant during the comparator operation. We will also only look for scaling relationships and see how transistor parameters contribute to the noise. This leads to considerably simpler expressions compared with more detailed analysis as found in [5].

Initialization Phase
The noise signal is now assumed to be approximated by the simple noise source across the source-drain of the input-stage transistors.

Simplify In this stage the clock switch at the bottom of the transconductor is on which activates the input stage. We will assume here that the capacitance of A is small compared with the full output load, $C_A \ll C_o$, and then the output transistor, M_3 looks like a CG stage. This is the opposite limit we looked at before when we were interested in maximum speed possible; here we look at the situation where we have an appreciable load at the output. For cases where the input stage is limited in size by the required speed and one uses a typical output load this is a reasonable assumption. We will also assume the input is itself a constant during this phase. Furthermore, we will assume the noise average over the timescales involved is

$$\langle i_n \rangle^2 = f_{BW,i}\, 4kT g_m \gamma$$

where we estimate the bandwidth, $f_{BW,i}$, to scale as 1 over a relevant timescale, here

$$f_{BW,i} \sim \frac{1}{\tau_i} \approx \frac{I_b}{V_t C_o}.$$

Solve We will look at this problem in the time domain. The output voltage is simply

$$C_o \frac{dv_o^n(t)}{dt} = -i_n(t).$$

We only know time average noise power $\sqrt{\int i_n^2(t)dt/T} = \langle i_n \rangle$. Therefore the solution to the differential equation can be estimated to be on average

$$\langle v_o^n(\tau_i) \rangle \sim \frac{\langle i_n \rangle}{C_o} \tau_i$$

$$\langle v_o^n(\tau_i) \rangle^2 \sim \frac{\langle i_n \rangle^2}{C_o^2} \tau_i^2 = \frac{4kT g_{m,1} \gamma}{C_o^2} \tau_i.$$

For the signal itself we find similarly, using $i(t) = g_m v_{in}$

$$v_o(\tau_i)^2 = \frac{g_m^2 v_{in}^2}{C_o^2} \tau_i^2.$$

The second transistor does not contribute to noise in this stage with these simplifications.

Regeneration Phase
For the regeneration phase we will again look at Figure 2.17.

Simplify We now simplify by assuming the drain of the input pair has reached ground so we simply have two cross-coupled inverters. The input transistors are thus

3.4 Comparator Circuit

disconnected from the output and do not contribute to the noise and not directly to the output during this phase.

Solve Due to the positive feedback nature of the cross-coupled pair we find again using simple scaling arguments that

$$\langle v_{n,o}(t)\rangle^2 \sim \frac{\langle i_n\rangle^2}{C_o^2}\tau_r^2 \exp\left(2\frac{t-\tau_i}{\tau_r}\right) = f_{BW,r}\frac{4kT\gamma\,(g_{m,p}+g_{m,n})}{C_o^2}\tau_r^2 \exp\left(2\frac{t-\tau_i}{\tau_r}\right)$$

$$f_{BW,r} \sim \frac{1}{\tau_r} = \frac{(g_{m,n}+g_{m,p})}{C_o}.$$

We get

$$\langle v_{n,o}(t)\rangle^2 \sim \frac{4kT\gamma\,(g_{m,p}+g_{m,n})}{C_o^2}\tau_r \exp\left(2\frac{t-\tau_i}{\tau_r}\right) = \frac{1}{C_o}4kT\gamma \exp\left(2\frac{t-\tau_i}{\tau_r}\right)$$

where we have used the timescale from equation (3.2) as the starting point for the exponential growth. The output noise voltage is highly time dependent, and the seed noise at the start of the regeneration phase will grow more than the noise voltage injected at later times.

Final Result
Putting all these together we find the output noise is

$$\langle (v_{n,o}^{final}(t))\rangle^2 \sim \frac{1}{C_o}4kT\gamma \exp\left(2\frac{t-\tau_i}{\tau_r}\right) + \frac{4kTg_{m,1}\gamma}{C_o^2}\tau_i \exp\left(2\frac{t-\tau_i}{\tau_r}\right).$$

The signal at the output is similarly

$$v_o^{final\,2} = \frac{g_{m,1}^2 v_{in}^2}{C_o^2}\tau_i^2 \exp\left(2\frac{t-\tau_i}{\tau_r}\right),$$

from which we see the gain

$$G^2 = \frac{g_{m,1}^2}{C_o^2}\tau_i^2 \exp\left(2\frac{t-\tau_i}{\tau_r}\right).$$

The input-referred noise is now

$$v_{n,i}^2 = \left(\frac{v_{n,o}^{final}(t)}{G}\right)^2 \sim \left(\frac{I_b}{V_t C_o}\right)^2\frac{C_o 4kT\gamma}{g_{m,1}^2} + \frac{I_b}{V_t C_o}\frac{4kT\gamma}{g_{m,1}}. \quad (3.5)$$

Verify The calculation here is a simplified version of a much more general discussion in, for example, [5], though we get similar information with significantly less effort. However, the full solution in [5] is certainly also valuable.

Evaluate This means that if we can maximize $g_{m,1}/I_b$ both terms will be small. Bear in mind the initial assumption that the input transistor is limited in size due to bandwidth limitations, which will limit what we can do. A more detailed analysis for the general case where the capacitance at the output of the transconductor is significant leads to similar conclusions.

3.5 Cascaded Amplifier Stages

This section concerns cascaded amplifier stages. In almost any modern electronic system there is more than one amplifier, and combining several together for an aggregate gain and linearity is a key skill. Here too a simplified estimation model can be constructed, as discussed in many books. We include such a model here more for the sake of completeness than in the hope of adding to the reader's knowledge.

Imagine the situation as in Figure 3.7. We have two amplifier stages where the noise and gain and linearity for both stages are marked in the figure. We will calculate the input-referred noise and linearity.

Simplify We simplify the picture by assuming the noise is not input-dependent and the gain has no phase shift. All entities are in power, as V^2/R.

Solve Let us first calculate the overall signal gain at the output. This is straightforward.
$$G = G_1 \cdot G_2.$$
The noise at the output is now
$$n_{out} = n_1 G + n_2 G_2.$$
The input-referred noise is simply this quantity divided by the gain
$$n_{in} = \frac{n_{out}}{G} = \frac{n_1 G + n_2 G_2}{G} = n_1 + \frac{n_2}{G_1}.$$
The total input-referred noise due to n_2 is reduced compared to n_1 by the gain of the first stage. Thus a high gain at the input will reduce the impact of noise at the subsequent stages. This is a useful property.

The linearity behaved differently, as we can see from Figure 3.7. At the output
$$P_o = \left(P_{in} G_1 + a_3 P_{in}^3\right) G_2 + b_3 (P_{in} G_1)^3,$$
where we have only kept terms up to third order. Divide by G to get the input-referred power finds

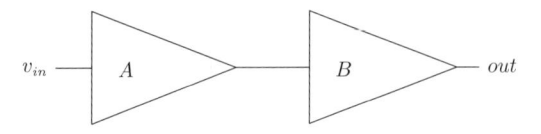

Figure 3.7 Two cascaded amplifiers.

3.5 Cascaded Amplifier Stages

$$P_o = \left(P_{in}G_1 + a_3 P_{in}^3\right)\frac{G_2}{G} + b_3(P_{in}G_1)^3 \frac{1}{G}.$$

We now see the non-linearity caused by the second stage is amplified by a factor $G_1^3/G = G_1^2/G_2$. In this case having a large gain upfront is detrimental to performance.

Verify These calculations and similar discussions can be found in [10], for example.

Evaluate The important thing to remember here is that for noise optimization it is convenient to have a large gain up front while for linearity optimization the gain should be small in the first stage. This inherent juxtaposition significantly complicates receiver design.

Design Examples

Example 3.1 Comparator Design
We will here discuss a comparator design based on the specifications detailed in Table 3.1. We will use this design in Chapter 7 when we design a full ADC.

Solution
With the timescaling rules discussed in the section "Comparator Analysis" it is fairly straightforward to size up the circuit. With different specifications one might use a different approach. Here we will

1. Size the input stage to be as large as maximum load specification allows. We know from the scaling rules we need the input stage to be as large as possible.
2. Size the cross-coupled pair to be large enough to meet the regeneration timescale specified.
3. Size the reset switches to be large enough to reset the nodes in roughly the initialization timescale.

The input stage should be three-unit transistors from Appendix A to meet the input load requirement. This will provide about 18 fF input capacitance, which is close to what is needed. From this size we know the output junction capacitance is about 6 fF. The cross-coupled pair should then be sized so that they are as large as possible without overly loading the input pair. The initialization timescale is perhaps five to six times shorter than the regeneration timescale, so for sizing concerns the regeneration timescale should be the focus. A cross-coupled transistor pair with $n_f = 10$ will provide

Table 3.1 Specification table for comparator design

Specification	Value	Comment
Input capacitance	<25 fF	
Output load	5 fF	Minimum size inverters
Regeneration timescale	5 ps	From system
Sampling period	50 ps	From system
Clock/reset rise time	10 ps	Assume as given

Table 3.2 Estimated and simulator optimized sizes for comparator decision circuits

Device	Size	Comment
M_1, thin oxide	W/L = 30 µm/27 n	
M_3, thin oxide	W/L = 10 µm/27 n	

Simulator optimization.

Device	Size	Comment
M_1, thin oxide	W/L = 30 µm/27 n	
M_3, thin oxide	W/L = 16 µm/27 n	

Table 3.3 Final size for comparator circuit including reset switches

Device	Size	Comment
M_1, thin oxide	W/L = 30 µm/27 n	
M_3, thin oxide	W/L = 16 µm/27 n	
Mreset, thin oxide	W/L = 16 µm/27 n	

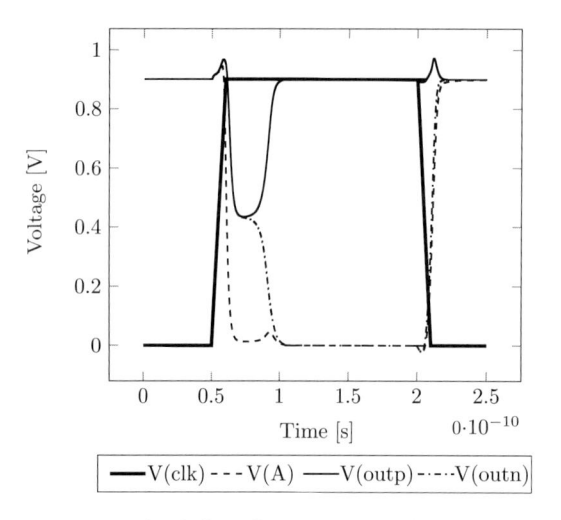

Figure 3.8 A simulation of a comparator with the sizing set by Table 3.3.

around $2 * 2C_{ox}$ capacitance as a load or around 32 fF with the transistor in Appendix A. We should start with this size of transistor, which will be around 5× the output load of the input pair (6 fF). For estimated and optimized parameters see Table 3.2.

Our initial estimate of the size is quite close to what we found with the simulator. The reset switches are finally sized so the output nodes are reset properly in roughly the initialization timescale. We find the final sizing in Table 3.3:

Figure 3.8 shows a simulation of the final sizing of the comparator.

Noise Estimation

The noise can be estimated from equation (3.5)

$$v_{n,rms}^{comp} \sim \sqrt{\frac{I_b}{V_t C_O} \frac{4kT\gamma}{g_{m,1}}} = \sqrt{\frac{3 \cdot 10^{-3}}{0.35 \cdot 25 \cdot 10^{-15}} \frac{4 \cdot 1.38 \cdot 10^{-23} \cdot 300 \cdot 2}{0.024}} \approx 0.7 \text{ mV}$$

Example 3.2 Amplifier with follower

This section discusses a common amplifier design with a low impedance output buffer. It will be used in Chapter 7 as part of a larger ADC design. The specifications in Table 3.4 is a result of a system evaluation of such an ADC.

Solution

Given the low output impedance and the low input capacitance it seems best to work with the topology shown in Figure 3.9.

The sizing is relatively straightforward. We will use the transistor size from Appendix A, which gives $m = 1$. The load will be set by the gain to be around 300 ohm, with a $g_m = 8$ mmho for the transconductor that will provide a gain of about 2.5.

Table 3.4 Specification table for amplifier with buffer

Specification	Value	Comment
Input capacitance	<10 fF	Driven by 10 ohm resistor but with 64 amplifiers in series, providing a time constant of 10 fF * 10 * 640 = 6 ps
Gain	>2	From system
Output load	20 fF	Differential input stage
Output impedance	≪100 ohms	From system
Output common mode	650 mV	From system

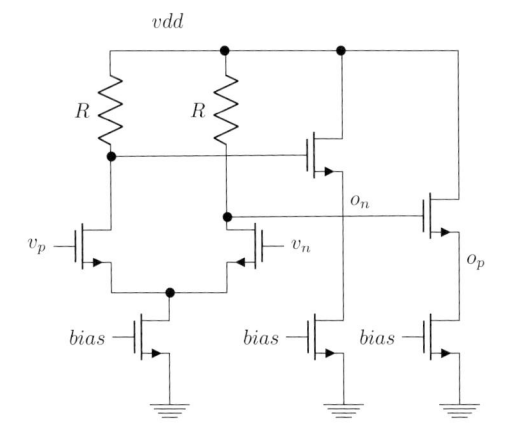

Figure 3.9 Amplifier with output buffer.

Table 3.5 Starting point which is also final sizes for amplifier

Device/Parameter	Value	Comment
M_1, M_2, M_3, M_4, channel width/length/nf	1 μm/30 nm/10	
Load resistor	300 ohm	
Input current	500 μA	Leaves a gain of 4 for main bias
M_4 channel width/length/nf	2 μm/250 n/10	Thick oxide device
M_5 channel width/length/nf	2 μm/250 n/40	Thick oxide device
M_6 channel width/length/nf	2 μm/250 n/20	Thick oxide device
Supply voltage	1.4 V	To get the output common mode correct

The bias current will be set by 2 mA from the bias of the unit transistor and the output follower will also be a unit size transistor since that has an output impedance of about 10 ohms. In summary the parameters will be as in Table 3.5.

For this case all the relevant properties have already been characterized and one would expect the simulation to line up closely to the estimated numbers. One thing to note is that the load resistance is about a third of the transistor output impedance. One cannot make it much larger and expect the corresponding benefit of gain increase.

3.6 Summary

We have looked at more complex transistor gain stages from an estimation analysis angle. We used the time description of the circuit equations when discussing comparators and we used several single pole analyses to get a handle on the timescales and noise transfer in a strong-arm type of comparator. We found that fairly simple arguments can be used to come close to solutions that are based on more detailed models. It is another example of the applicability of estimation analysis where we consider the core behavior of the system and try to capture it with simple modeling. The steps simplify, solve, verify and evaluate were used time and time again and we used the transistor size parameters suggested by simple modeling as a starting point for the simulator fine-tuning. Finally we looked at cascaded amplifier stages where well-known relationships between noise and linearity were derived following the estimation analysis steps.

3.7 Exercises

1. Redo the noise analysis for the five transistor circuit where the bias transistor is replaced by a bias resistor. Do not rush into the calculations. Instead, simplify and estimate.

2. Redo the comparator noise analysis where the capacitance at the drain of the input pair cannot be neglected. Simplify, Solve, Verify, Evaluate!
3. Improve the cascaded feedback amplifier model by assuming the amplifier has a finite bandwidth. (Treat it as a single pole at $\omega_{3\,\text{dB}}$.)

3.8 References

[1] D. A. Johns and K. Martin, *Analog Integrated Circuit Design*, Hoboken, NJ: Wiley, 1996.

[2] R. Gray, J. Hurst, S. Lewis, and R. Meyer, *Analysis and Design of Analog Integrated Circuits*, 5th edn., New York: Wiley, 2009.

[3] B. Razavi, *Design of Analog CMOS Integrated Circuits*, 2nd edn., New York: McGraw-Hill, 2016.

[4] F. Maloberti, *Analog Design for CMOS VLSI Systems*, Dordrecht: Kluwer Academic, 2003.

[5] J. Kim, B. S. Leibowitz, J. Ren, and C.J. Madden, "Simulation and Analysis of Random Decision Errors in Clocked Comparator," *IEEE Transactions on Circuits and Systems*, Vol. 56, No. 8, pp. 1844–1857, 2005.

[6] R. Vinoth and S. Ramasamy, "Design and Implementation of High Speed Latched Comparator Using gm/Id Sizing Method," *ICTACT Journal on Microelectronics*, Vol. 2, No. 04, pp. 300–304, 2017.

[7] B. Razavi, "The StrongARM Latch," *IEEE Solid-State Circuits Magazine*, Spring 2015.

[8] F. Maloberti, *Data Converters*, Dordrecht: Springer, 2008.

[9] R. van de Plassche, *CMOS Integrated Analog-to-Digital and Digital-to-Analog Converters*, 2nd edn., Dordrecht: Kluwer Academic, 2003.

[10] J. Rogers and C. Plett, *Radio Frequency Integrated Circuit Design*, Norwood, MA: Artech House, 2003.

4 Electromagnetism: Fundamentals

> **Learning Objectives**
>
> - Maxwell's equations in source form
> - Using estimation analysis in connection with Maxwell's equations
> - Duality of capacitance vs inductance
> - One to three-dimensional solutions to Maxwell's equations relevant to integrated circuit designers
> - Current distributions in various situations in order to estimate inductance

4.1 Introduction

This chapter discusses the basis of electromagnetism in terms of Maxwell's equations. This topic that has been studied extensively and there are many great books that discuss its various aspects: see [1–14] for a small selection. Items [2, 5] focus on microwave aspects of the theory. Items [4, 8, 12] are standard physics graduate student texts. A more recent treatment is, for example [13, 14], which showcases engineering aspects of electromagnetism. We will follow the presentation in [1, 2] fairly closely. The intention in this chapter is to be self-consistent and in that spirit we will present common solution techniques found in the literature. The estimation techniques we discuss in this book will be heavily applied toward the end of the chapter where the concepts of capacitance and inductance are introduced. It is assumed that the reader has encountered electromagnetism before in elementary classes and we will not discuss the basic discoveries and the history that led to the remarkable formulation of the fundamental equations by Maxwell in a series of papers around 1865. The history of this development is a fascinating read and a great example of how science evolves [3].

We will start with a brief discussion of Maxwell's equations and show how to reformulate them to be suitable in various situations encountered in integrated circuit design. Common solution techniques and handling of boundary conditions are presented next. Thereafter we will discuss the important concept of energy and power relating to the electromagnetic fields. These concepts will naturally lead to the definitions of capacitance and inductance both in general and for circuit theory. We will show that the concepts are naturally very similar, or dual, and we attempt to dispel some of the mystery that sometimes surrounds these phenomena. The chapter wraps up with a handful of examples where the estimation analysis technique is applied to calculate capacitance, inductance, skin effect, and other such effects. Most of these examples will start directly from Maxwell's equations.

4.2 Maxwell's Equations

This section presents Maxwell's equations and we will follow the general outline presented in [1, 2]. Maxwell's work was based on a large body of empirical and theoretical knowledge developed by Gauss, Ampere, Faraday, and others.

It is assumed that the reader has some familiarity with Maxwell's equations and the history leading to their discovery. Here we will simply state them and highlight some of the historical events that surrounds them. The equations will be presented in their differential form. We believe most readers are familiar with the MKS or SI system of units and we will use them throughout the book.

With this we have for the equations:

$$\nabla \times \boldsymbol{H} = \frac{\partial \boldsymbol{D}}{\partial t} + \boldsymbol{J}. \tag{4.1}$$

$$\nabla \cdot \boldsymbol{D} = \rho. \tag{4.2}$$

$$\nabla \times \boldsymbol{E} = -\frac{\partial \boldsymbol{B}}{\partial t}. \tag{4.3}$$

$$\nabla \cdot \boldsymbol{B} = 0. \tag{4.4}$$

We have the quantities defined as:
H is the magnetic field in amperes per meters [A/m]
D is the electric flux density, in coulombs per meter squared [coul/m^2]
J is the electric current density in amperes per meter squared [A/m^2]
ρ is the electric charge density in coulombs per meter cubed [coul/m^3]
E is the electric field in volts per meter [V/m]
B is the magnetic flux density in webers per meter squared [Wb/m^2]

The fields and their corresponding fluxes are related by the constitutional equations:

$$\boldsymbol{D} = \epsilon \boldsymbol{E}. \tag{4.5}$$

$$\boldsymbol{B} = \mu \boldsymbol{H} \tag{4.6}$$

where
ϵ is the permittivity in farad per meter [F/m]
μ is the permeability in henry per meter [H/m]

The factors ϵ, μ are matrices in general and dependent on position. Throughout the book we will assume them to be scalar functions that are occasionally dependent on position.

Vector Potential and Elementary Gauge Theory

Having established Maxwell's equations we can now take note of some interesting properties. For one, there are no magnetic charges, hence $\nabla \cdot \boldsymbol{B} = 0$. This means that instead of using $\boldsymbol{B}$ we can define another entity $\boldsymbol{A}$ by using the fact that for functions that are smooth, all derivatives exists and are continuous, we have the vector identity $\nabla \cdot (\nabla \times \boldsymbol{A}) = 0$. This has important implications. If

$$\boldsymbol{B} = \nabla \times \boldsymbol{A} \tag{4.7}$$

then equation (4.4) is automatically fulfilled. Substituting this into (4.3) we get

$$\nabla \times \boldsymbol{E} = -\frac{\partial \boldsymbol{B}}{\partial t} = -\frac{\partial (\nabla \times \boldsymbol{A})}{\partial t} = -\nabla \times \frac{\partial \boldsymbol{A}}{\partial t}$$

We now use the vector identity: $\nabla \times \nabla \varphi \equiv 0$, where $\varphi(\boldsymbol{x},t)$ is any smooth function of coordinates $\boldsymbol{x}$ and time t, to integrate $\boldsymbol{E}$ and we find

$$\boldsymbol{E} = -\nabla \varphi - \frac{\partial \boldsymbol{A}}{\partial t} \tag{4.8}$$

The vector field $\boldsymbol{A}$ is commonly referred to as the vector potential field and scalar φ is known as the potential field, or voltage field. Together they are called gauge potentials in physics literature.

The equation for the E-field should be familiar to most readers with the possible exception of the last term. Equation (4.8) is simply the normal elementary textbook definition of the electric field as a gradient of a voltage but with an additional time-derivative (dynamic) term. It means we can have an electric field without a voltage drop in a dynamic situation. With the potential fields we can write Maxwell's equations to be a set of equations for φ and $\boldsymbol{A}$. Let us rewrite equation (4.1):

$$\nabla \times \boldsymbol{H} = \nabla \times \frac{\nabla \times \boldsymbol{A}}{\mu} = \frac{\partial \boldsymbol{D}}{\partial t} + \boldsymbol{J} = \frac{\partial}{\partial t}\epsilon\left(-\nabla \varphi - \frac{\partial \boldsymbol{A}}{\partial t}\right) + \boldsymbol{J}$$

or

$$\nabla \times \frac{\nabla \times \boldsymbol{A}}{\mu} = \frac{\partial}{\partial t}\epsilon\left(-\nabla \varphi - \frac{\partial \boldsymbol{A}}{\partial t}\right) + \boldsymbol{J} \tag{4.9}$$

and from equations (4.2) and (4.5) we find

$$\nabla \cdot \epsilon\left(-\nabla \varphi - \frac{\partial \boldsymbol{A}}{\partial t}\right) = \rho. \tag{4.10}$$

We now have another version of Maxwell's equations through equations (4.7)–(4.10). As the reader will have noticed, the potentials, $\boldsymbol{A}, \varphi$ are not uniquely defined. One can, for instance, add a term $\sim \nabla f$ where f is some function to $\boldsymbol{A}$ and the $\boldsymbol{B}$ field is not affected ($\nabla \times \nabla f \equiv 0$). This freedom in the choice of the potentials is known as gauge invariance. Let us look at this in some more detail (compare with [15] for a similar argument).

Let $\gamma(\mathbf{x}, t)$ be an arbitrary scalar field. Then let us change the gauge potentials according to the following transformation

$$\varphi \to \varphi - \frac{\partial}{\partial t}\gamma \quad \mathbf{A} \to \mathbf{A} + \nabla\gamma. \tag{4.11}$$

Now

$$\mathbf{B} \to \nabla \times (\mathbf{A} + \nabla\gamma) = \nabla \times \mathbf{A}$$

and

$$\mathbf{E} \to -\nabla\left(\varphi - \frac{\partial}{\partial t}\gamma\right) - \frac{\partial}{\partial t}(\mathbf{A} + \nabla\gamma) = -\nabla\varphi - \frac{\partial \mathbf{A}}{\partial t}.$$

Both fields are unchanged! The transformation in equation (4.11) is known as a gauge transformation, and since the fields are unchanged under this transformation we speak of gauge symmetry. We see that a physical system is described by a whole family of gauge potentials that differ by a gauge transformation. By picking a particular set of gauge potentials we are making a gauge choice. The above might seem a trivial matter but it has profound importance in theoretical physics. The interested reader is highly encouraged to refer to the literature on this matter.

We will now show that we can always find a solution that satisfies

$$\nabla \cdot \mathbf{A} + \mu\epsilon \frac{\partial}{\partial t}\varphi = 0 \tag{4.12}$$

by making an appropriate gauge choice. Assume we have a particular solution $\mathbf{A}', \varphi'$. We look for a particular gauge transformation that will satisfy (4.12). We have using (4.11)

$$\nabla \cdot (\mathbf{A}' + \nabla\gamma) + \mu\epsilon \frac{\partial}{\partial t}\left(\varphi' - \frac{\partial}{\partial t}\gamma\right) = \nabla \cdot \mathbf{A}' + \mu\epsilon \frac{\partial}{\partial t}\varphi' + \Delta\gamma - \frac{\partial^2}{\partial t^2}\gamma = 0$$

or

$$\Delta\gamma - \frac{\partial^2}{\partial t^2}\gamma = -\left(\nabla \cdot \mathbf{A}' + \mu\epsilon \frac{\partial}{\partial t}\varphi'\right).$$

The right-hand side is the known solution which acts as a source term for a wave equation which we can solve for γ. This way we can always find gauge potentials that satisfy (4.12). We do not have to find γ explicitly, but we can use this calculation as a motivation to use (4.12) as an additional requirement on $\mathbf{A}, \varphi$. Equation (4.12) is known as the Lorenz gauge. Another typical choice is the Coulomb gauge

$$\nabla \cdot \mathbf{A} = 0. \tag{4.13}$$

The Lorenz gauge is typically used for situations where the wavelength is comparable to the physical sizes. It is standard in microwave theory and antenna theory for obvious reasons. For integrated circuits one can often get by with the Coulomb gauge which corresponds to the Lorenz gauge in the long-wavelength approximation.

Maxwell's Equations in Terms of External Sources

In electrical engineering it is natural to think of currents and charges as being impressed on an electric system through voltage or current sources. These impressed entities will then give rise to electromagnetic fields. We will now write Maxwell's equation in terms of such impressed currents and charges.

The current can be divided into two parts. A conduction component,

$$J_c = \sigma E \quad \text{(ohm's law)} \tag{4.14}$$

and an impressed current, J_i. The charge can be divided the same way. By taking the divergence of equation (4.1) we recover the continuity equation:

$$\frac{\partial \rho}{\partial t} = -\nabla \cdot J \tag{4.15}$$

which relates the charge to the current. The addition of the time derivative of the D-field to Ampere's law ($\nabla \times H = J$) was Maxwell's famous generalization that created a self-consistent description of the field equations. For the current we have

$$J = J_c + J_i = \sigma E + J_i$$
$$\rho = \rho_c + \rho_i$$
$$\nabla \cdot (\sigma E) = -\frac{\partial \rho_c}{\partial t} \quad \text{(continuity equation for the conduction component)}.$$

Putting this together we get

$$\nabla \times H = \frac{\partial \epsilon E}{\partial t} + J = \frac{\partial \epsilon E}{\partial t} + J_c + J_i = \frac{\partial \epsilon E}{\partial t} + \sigma E + J_i$$
$$\nabla \times H - \frac{\partial \epsilon E}{\partial t} - \sigma E = J_i \tag{4.16}$$

$$\nabla \cdot (\epsilon E) = \rho = \rho_c + \rho_i$$
$$\nabla \cdot (\epsilon E) - \rho_c = \rho_i. \tag{4.17}$$

These equations show the fields as a result of external source currents and charges.

Full-Wave Approximation – Single Frequency Tone Formulation

In this book we will generally look at these equations not as a function of time but as a function of frequency. We get there by simply assuming the time dependence scales as $e^{j\omega t}$ where we follow the convention in most engineering books. By doing this we get

$$\nabla \times H = j\omega \epsilon E + \sigma E + J_i = j\omega \epsilon \left(1 + \frac{\sigma}{j\omega\epsilon}\right) E + J_i. \tag{4.18}$$

From the continuity equation for the conduction component we get

$$\nabla \cdot (\sigma E) = -\frac{\partial \rho_c}{\partial t} = -j\omega \rho_c \tag{4.19}$$

This together with the charge equation (4.17) gives

4.2 Maxwell's Equations

$$\nabla \cdot (\epsilon \mathbf{E}) = \rho = \rho_c + \rho_i = \frac{\nabla \cdot (\sigma \mathbf{E})}{-j\omega} + \rho_i \rightarrow \nabla \cdot (\epsilon \mathbf{E}) + \frac{\nabla \cdot (\sigma \mathbf{E})}{j\omega} = \rho_i \quad (4.20)$$

We can now define an effective permittivity

$$\epsilon' = \epsilon \left(1 + \frac{\sigma}{j\omega\epsilon}\right) \quad (4.21)$$

and we have

$$\nabla \times \mathbf{H} = j\omega \, \epsilon' \, \mathbf{E} + \mathbf{J}_i \quad (4.22)$$

$$\nabla \cdot (\epsilon' \mathbf{E}) = \rho_i. \quad (4.23)$$

Using equations (4.6)–(4.8) we get

$$\nabla \cdot \mathbf{B} = \nabla \cdot \nabla \times \mathbf{A} = 0$$
$$\nabla \times \mathbf{E} = \nabla \times (-\nabla \varphi - j\omega \mathbf{A}) = -j\omega \nabla \times \mathbf{A} = -j\omega \mathbf{B}$$
$$\nabla \times \mathbf{H} = \nabla \times \frac{\mathbf{B}}{\mu} = \frac{1}{\mu} \nabla \times \nabla \times \mathbf{A} = \frac{1}{\mu} \left(\nabla (\nabla \cdot \mathbf{A}) - \nabla^2 \mathbf{A}\right) = j\omega \epsilon' (-\nabla \varphi - j\omega \mathbf{A}) + \mathbf{J}_i.$$

We use the Lorenz gauge which in frequency domain looks like

$$\nabla \cdot \mathbf{A} + j\epsilon' \mu \omega \, \varphi = 0 \quad (4.24)$$

and get

$$\nabla (\nabla \cdot \mathbf{A}) - \nabla^2 \mathbf{A} = \nabla (-j\epsilon' \mu \omega \, \varphi) - \nabla^2 \mathbf{A} = j\omega \mu \, \epsilon' (-\nabla \varphi - j\omega \mathbf{A}) + \mu \mathbf{J}_i$$

After rewriting we find

$$\nabla^2 \mathbf{A} + \omega^2 \, \mu \, \epsilon' \mathbf{A} = -\mu \mathbf{J}_i \quad (4.25)$$

We also have from equation (4.23)

$$\nabla \cdot (\epsilon' \mathbf{E}) = \epsilon' \nabla \cdot \mathbf{E} = \epsilon' \nabla \cdot (-\nabla \varphi - j\omega \mathbf{A}) = -\epsilon' \Delta \varphi + j\omega \epsilon' \, j\epsilon' \mu \omega \, \varphi = -\epsilon' \Delta \varphi - \omega^2 \epsilon' \epsilon' \mu \varphi = \rho_i$$

After rewriting we find

$$\Delta \varphi + \omega^2 \epsilon' \mu \varphi = -\frac{\rho_i}{\epsilon'} \quad (4.26)$$

Equations (4.25) and (4.26) are Maxwell's equation in yet another form. Knowing $\mathbf{A}, \varphi$ will give us $\mathbf{B}$ and $\mathbf{E}$ through equations (4.7) and (4.8). We will use these in numerous examples in this and the following chapters.

Long Wavelength Approximation
In the long wavelength approximation, $\lambda \gg l$, where $\lambda = 2\pi c/\omega$ is the wavelength and l is a length scale of the model, we find the second term on the left-hand side of (4.25), (4.26) disappears and we are left with:

$$\nabla^2 \mathbf{A} = -\mu \mathbf{J}_i. \tag{4.27}$$

We can also write this directly from the fields as

$$\nabla \times \mathbf{H} = \mu \mathbf{J}_i \quad \text{(Ampere's law)}, \tag{4.28}$$

where (4.27) follows if we use gauge $\nabla \cdot \mathbf{A} = 0$ and

$$\Delta \varphi = -\frac{\rho_i}{\epsilon'}. \tag{4.29}$$

(This follows also from $\nabla \cdot (\epsilon' \mathbf{E}) = \rho_i$ if we use gauge $\nabla \cdot \mathbf{A} = 0$)

For integrated circuits the long wavelength approximation is often appropriate since the dimensions are much smaller than any wavelength.

Solutions to Maxwell's Equations

Note the equations are of the same form where the only difference is the vector form for the vector potential and scalar from for the voltage field equation. The general solutions are known for certain sources. Here we will almost exclusively look at cases where the sources are Dirac delta functions of one sort or other. The specific solution will depend on the boundary conditions and most of the time will be spent establishing those.

For completeness, in this section we will discuss a common approach to solve wave style equations such as Maxwell's. The first subsection covers the general solutions and we will discuss how to handle the all-important boundary conditions in the following subsection. We follow the presentation given in [1].

General Solution

We will start discussing the general solution of the one-dimensional case and follow with the two- and three-dimensional versions in the following subsections. There is a rich literature describing these methods and also a list of references at the end of the chapter.

1D – Solution

Let us consider an equation similar to (4.26) in free space

$$\frac{d^2 \varphi(x)}{dx^2} + k^2 \varphi(x) = -\delta(x - x_0). \tag{4.30}$$

This is known as Helmholz's equation in one-dimensional free space subject to the boundary condition at infinity, $\varphi(\pm\infty) = 0$. The response at x is due to the delta source at x_0. Let us consider the homogeneous equation

$$\frac{d^2 \varphi(x)}{dx^2} + k^2 \varphi(x) = 0$$

This is the same as (4.30) when $x \neq x_0$. The solution to this equation that satisfies the boundary conditions at infinity is

$$\varphi(x) = \begin{cases} Ae^{jkx} & x > x_0 \\ Be^{-jkx} & x < x_0. \end{cases}$$

The unknown constants A, B can be determined by the boundary condition at $x = x_0 \pm \Delta$ where Δ denotes an infinitesimally small interval. Integrating (4.30) from $x = x_0 - \Delta$ to $x = x_0 + \Delta$ we find

$$\left[\frac{d\varphi}{dx}\right]_{x=x_0-\Delta}^{x=x_0+\Delta} + \int_{x=x_0-\Delta}^{x=x_0+\Delta} k^2 \varphi(x) dx = -1.$$

Since $\varphi(x)$ is continuous the last term on the left-hand side disappears when $\Delta \to 0$.

We get

$$\begin{cases} \left[\frac{d\varphi}{dx}\right]_{x=x_0-\Delta}^{x=x_0+\Delta} = -1 \\ \varphi(x+\Delta) = \varphi(x-\Delta) \end{cases}$$
$$\begin{cases} ik\left(Ae^{jkx} + Be^{-jkx}\right) = -1 \\ Ae^{jkx} = Be^{-jkx}. \end{cases}$$

Solving for A and B gives

$$\varphi(x) = \begin{cases} \frac{j}{2k} e^{jk(x-x_0)} & x > x_0 \\ \frac{j}{2k} e^{-jk(x-x_0)} & x < x_0 \end{cases}$$
$$= \frac{i}{2k} e^{jk|x-x_0|}. \tag{4.31}$$

Long Wavelength Approximation

When $kx_0 \ll 1$ (long wavelength approximation) we find

$$\varphi_{lw}(x) = \begin{cases} \frac{j}{2k}(1 + jk(x-x_0)) & x > x_0 \\ \frac{j}{2k}(1 - jk(x-x_0)) & x < x_0 \end{cases}$$
$$= \frac{j}{2k}(1 + j|k(x-x_0)|) = \frac{j}{2k} - \frac{|x-x_0|}{2}.$$

Since in the long wavelength approximation Helmholz equation reduces to Poisson equation where φ_{lw} is defined with an arbitrary constant factor, we can simply relabel the constant term for φ_{lw} and end up with

$$\varphi_{lw}(x) = -\frac{|x-x_0|}{2} + C. \tag{4.32}$$

Let us verify by examining the Poisson equation

$$\frac{d^2\varphi_{lw}(x)}{dx^2} = -\delta(x - x_0).$$

When $x \neq x_0$ we see $\frac{\partial^2\varphi_{lw}(x)}{\partial x^2} \equiv 0$. Let us integrate around the singularity as we did earlier

$$\int_{x_0-\Delta}^{x_0+\Delta} \frac{\partial^2\varphi_{lw}(x)}{\partial x^2} dx = -\int_{x_0-\Delta}^{x_0+\Delta} \delta(x - x_0) dx$$

$$\left[\frac{d\varphi_{lw}(x)}{dx}\right]_{x_0-\Delta}^{x_0+\Delta} = -\frac{1}{2} - \frac{1}{2} = -1 = r.h.s.$$

Indeed, in one dimension and the long wavelength approximation (4.32) solves the Poisson equation.

2D – Solution

In two dimensions equation (4.30) becomes

$$\frac{\partial^2\varphi(x, y)}{\partial x^2} + \frac{\partial^2\varphi(x, y)}{\partial y^2} + k^2 \varphi(x, y) = -\delta(y - y_0)\delta(x - x_0) \quad (4.33)$$

We can now use the Fourier transform

$$\varphi(x, y) = \int_{-\infty}^{\infty} \tilde{\varphi} e^{j\beta(x-x_0)} d\beta.$$

This gives

$$\frac{\partial^2\varphi(\beta, y)}{\partial y^2} - \beta^2 \tilde{\varphi}(\beta, y) + k^2 \tilde{\varphi}(\beta, y) = -\delta(y - y_0)$$

The solution to this equation is the one-dimensional free-space Green's function

$$\tilde{\varphi}(\beta, y) = \frac{j}{2\kappa} e^{j\kappa|y-y_0|}$$

where $\kappa = \sqrt{k^2 - \beta^2}$. We find

$$\varphi(x, y) = \int_{-\infty}^{\infty} \frac{j}{2\kappa} e^{j\kappa|y-y_0|} e^{j\beta(x-x_0)} d\beta = \frac{j}{4} H_0^{(1)}\left(k \sqrt{(x-x_0)^2 + (y-y_0)^2}\right)$$

where the last equality relates the Hankel function of order 0, $H_0^{(1)}$ to the integral. We see the solution is simply a composition of a continuous spectrum of plane waves.

Long Wavelength Approximation

The two-dimensional solution in the long wavelength approximation can be found using similar techniques as in the one-dimensional case we noted earlier. Here we will show the solution for the special case of cylindrical symmetry which we will take advantage of later in this chapter when we discuss inductance and current elements.

Helmholz equation in cylindrical symmetry becomes, with a delta function at $x = 0$,

$$\nabla \cdot \nabla \varphi = C\delta(\boldsymbol{x}). \qquad (4.34)$$

Outside of $\boldsymbol{x} = 0$ we have

$$\nabla \cdot \nabla \varphi = \frac{1}{r}\frac{\partial}{\partial r} r \frac{\partial \varphi(r,\theta)}{\partial r} + \frac{1}{r^2}\frac{\partial^2 \varphi(r,\theta)}{\partial \theta^2} = 0.$$

To further simplify we will assume there is no θ-dependence and we find

$$\frac{1}{r}\frac{d}{dr} r \frac{d\varphi(r)}{dr} = \frac{d^2\varphi(r)}{dr^2} + \frac{1}{r}\frac{d\varphi(r)}{dr} = 0.$$

This has the general solution

$$\varphi(r) = D \ln r + B.$$

To find out the constants we need to integrate (4.34) around $\boldsymbol{x} = 0$. Let us choose a sphere centered at $\boldsymbol{x} = 0$ with radius Δ as an integration volume. We find for the left-hand side using the divergence theorem (see Appendix B)

$$\int \nabla \cdot \nabla \varphi dV = \int \nabla \varphi \cdot \frac{\boldsymbol{r}}{r} da = \int \frac{d\varphi}{dr} r d\theta = \frac{d\varphi}{dr} r 2\pi = 2\pi D.$$

The right-hand side of (4.35) becomes, as before, C. Putting all this together we have

$$2\pi D = C \rightarrow D = \frac{C}{2\pi}.$$

We have for the long wavelength solution to the Helmholz equation in two dimensions:

$$\varphi(r) = \frac{C}{2\pi} \ln r + B. \qquad (4.35)$$

3D – Solution

We finally present the 3D solution. We will use it in Chapter 6. Equation (4.30) becomes with a source at $\boldsymbol{r} = \boldsymbol{r}_0$

$$\frac{\partial^2 \varphi(x,y,z)}{\partial x^2} + \frac{\partial^2 \varphi(x,y,z)}{\partial y^2} + \frac{\partial^2 \varphi(x,y,z)}{\partial z^2} + k^2 \varphi(x,y,z) = -\delta(\boldsymbol{r} - \boldsymbol{r}_0).$$

Let us first do a change of variables $\boldsymbol{\rho} = \boldsymbol{r} - \boldsymbol{r}_0$. We find we have created a spherically symmetric model. By realizing

$$\int \delta(\boldsymbol{r} - \boldsymbol{r}_0) dV = \int \delta(\boldsymbol{\rho}) 4\pi \rho^2 d\rho,$$

we find

$$\delta(\boldsymbol{r} - \boldsymbol{r}_0) = \frac{\delta(\rho)}{4\pi \rho^2}.$$

By substituting

$$\varphi(\rho) = \frac{u(\rho)}{\rho},$$

we find for the Helmholz equation

$$\left(\nabla^2 + k^2\right)u(\rho) = -\frac{\delta(\rho)}{4\pi\rho}.$$

The solution for $u(\rho) = Ae^{jk\rho}$ and we find

$$\varphi(\rho) = \frac{Ae^{jk\rho}}{\rho}.$$

The boundary condition at $\rho = 0$ must be used to determine A. The Helmholz equation is

$$\left(\nabla^2 + k^2\right)\varphi(\rho) = -\frac{\delta(\rho)}{4\pi\rho^2}.$$

After integrating over a small volume we find, as with the one-dimensional case,

$$\varphi(0+) = \varphi(0-),$$
$$\int \nabla^2 \varphi(\rho) dV = -1.$$

After applying the divergence theorem we get

$$\int \nabla^2 \varphi(\rho) dV = \oint \nabla \varphi(\rho) \cdot ds = A \left(\frac{jke^{jk\rho}\rho - e^{jk\rho}}{\rho^2} 4\pi\rho^2\right)_{\rho \to 0} = -4\pi A.$$

So

$$A = \frac{1}{4\pi}.$$

The general solution is

$$\varphi(x,y,z) = \frac{1}{4\pi|\mathbf{r} - \mathbf{r}_0|} e^{j\boldsymbol{\kappa} \cdot (\mathbf{r} - \mathbf{r}_0)} + \frac{1}{4\pi|\mathbf{r} - \mathbf{r}_0|} e^{-j\boldsymbol{\kappa} \cdot (\mathbf{r} - \mathbf{r}_0)}, \quad \kappa^2 = \kappa_x^2 + \kappa_y^2 + \kappa_z^2 = \omega^2 \epsilon' \mu.$$

Similarly, for the vector potential we get

$$\mathbf{A}(\mathbf{r}) = \frac{\mu}{4\pi} \int \frac{\mathbf{J}_i(\mathbf{r}')}{|\mathbf{r} - \mathbf{r}'|} \left(e^{j\boldsymbol{\kappa} \cdot (\mathbf{r} - \mathbf{r}')} + e^{-j\boldsymbol{\kappa} \cdot (\mathbf{r} - \mathbf{r}')}\right) d\mathbf{r}'.$$

Long Wavelength Approximation
The solution to the long wavelength equations (4.27) and (4.29) in three dimensions are well known. We see the equations are essentially identical and the general solution that vanishes at infinity is

$$\mathbf{A}(\mathbf{r}) = \frac{\mu}{4\pi} \int \frac{\mathbf{J}_i(\mathbf{r}')}{|\mathbf{r} - \mathbf{r}'|} d\mathbf{r}'. \tag{4.36}$$

The solution for φ is similarly

$$\varphi(\mathbf{r}) = \frac{1}{4\pi\epsilon'} \int \frac{\rho_i(\mathbf{r}')}{|\mathbf{r} - \mathbf{r}'|} d\mathbf{r}'. \tag{4.37}$$

For a point charge at the origin, $\rho_i(\mathbf{r}') = q\delta(\mathbf{r}')$ we find

$$\varphi(\mathbf{r}) = \frac{q}{4\pi\epsilon' r}.$$

which is the familiar electrostatic potential from a point charge.

Boundary Conditions

Identifying the relevant equations and how to solve them is a helpful exercise. It is often the easiest part of any investigation. The real problem comes when taking into account what happens at the boundaries, either in time and/or in space. To the novice, the opposite often appears true. To help us there are many great examples in the literature on how to handle the boundaries, and here we will go through the basic methods and leave it to the reader to explore more if needed.

Fundamentally, what one does is to put a "pill box" at the boundary that extends a little into each material with a large surface area. In Figure 4.1 it extends $\varepsilon/2$ into each region. The volume is thus infinitesimal, while the area is macroscopic. They key thing to note here is that the equations are still valid in this volume and to find out what the boundary conditions are one simply integrates the equations over the small volume. For some entities, portions of the equation will be proportional to the volume and thus small while other entities will be proportional to the area and thus large. We will go through Maxwell's equations with this method to show explicitly how the conditions work out.

Let us look at Ampere's law (4.28)

$$\nabla \times \mathbf{H} = \mathbf{J}.$$

At the boundary between two media we put a small pill box of height ε which is much smaller than any other dimension in the problem. We integrate Ampere's law over this volume

$$\int \nabla \times \mathbf{H} \, dV = \int \mathbf{J} \, dV.$$

We can here use Stoke's theorem for the left-hand side and denoting by $\mathbf{A}_{rea} = A_{rea}\mathbf{n}$ where A_{rea} is the area and $\mathbf{n}$ is the outward normal to the area segment

$$\int \nabla \times \mathbf{H} \, dV = \oint \mathbf{H} \times \mathbf{A}_{rea} \, dA_{rea} = A_{rea}(H_{t,+} - H_{t,-}) + \varepsilon H_n \to A_{rea}(H_{t,+} - H_{t,-}), \varepsilon \to 0.$$

Volume Current

For the right-hand side without delta functions, which we call volume current, we get

$$\int \mathbf{J} \, dV \sim \varepsilon \to 0 \quad \text{when} \quad \varepsilon \to 0.$$

Putting it all together we get at the boundary

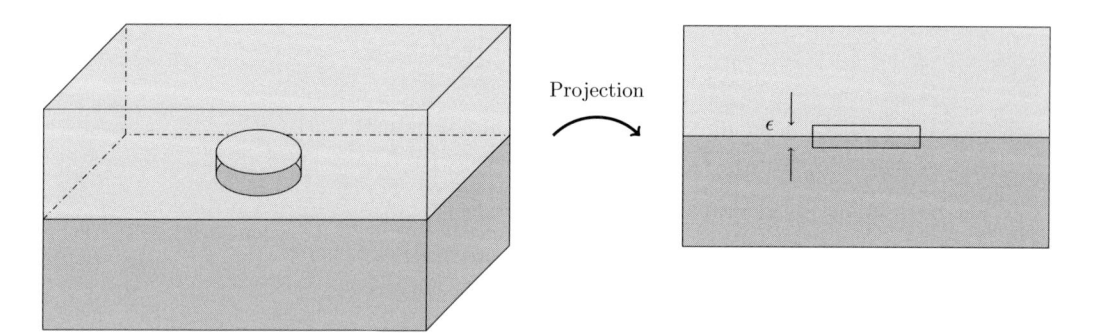

Figure 4.1 Boundary conditions pill box.

$$\int \nabla \times \boldsymbol{H} \, dV = A_{rea}(H_{t,+} - H_{t,-}) = \int \boldsymbol{J} \, dV = 0.$$

or

$$H_{t,+} = H_{t,-}. \qquad (4.38)$$

Surface Current

We also see when we have a delta function on the right-hand side, a surface current,

$$\int \boldsymbol{J} \, dV = \int J_s \delta(y) dV = J_s A_{rea} \quad \text{when} \quad \varepsilon \to 0.$$

Putting this together with the left-hand side we get

$$A_{rea}(H_{t,+} - H_{t,-}) = J_s A_{rea}$$

or

$$(H_{t,+} - H_{t,-}) = J_s \qquad (4.39)$$

Similarly, we can use the charge equation (4.23)

$$\nabla \cdot (\epsilon' \boldsymbol{E}) = \rho_i.$$

We integrate over volume

$$\int \nabla \cdot (\epsilon' \boldsymbol{E}) dV = \int \rho_i \, dV$$

For the left-hand side we use Gauss' law

$$\int \nabla \cdot (\epsilon' \boldsymbol{E}) dV = \oint \epsilon' \boldsymbol{E} \cdot \boldsymbol{n} \, dA_{rea} = A_{rea}\left(\epsilon'_+ E_{n+} - \epsilon'_- E_{n-}\right)$$

$$+ \varepsilon \ldots \to A_{rea}\left(\epsilon'_+ E_{n+} - \epsilon'_- E_{n-}\right), \varepsilon \to 0$$

4.2 Maxwell's Equations

Volume Charge
For the case of volume charge, the right-hand side is now treated the same way as for the volume current earlier so we end up with

$$\epsilon'_+ E_{n+} = \epsilon'_- E_{n-} \tag{4.40}$$

Finally, for the case of surface charges we have

Surface Charge

$$\epsilon'_+ E_{n+} - \epsilon'_- E_{n-} = \frac{\rho_s}{A_{rea}} \tag{4.41}$$

This way of treating the boundary is standard and can be found in many textbooks.

Field Energy Definitions

Let us look at the concept of energy in these fields. It is outside the scope of this book to fully derive this here but we will make some plausible arguments as to their validity: see [4] for details. Let us start with a static electric field where there are no currents.

Electric Field Energy
We assume for simplicity there is one conductor at a constant voltage φ. If we have an infinitesimal charge $\delta\rho$ moving from infinity to the conductor we will need to apply an energy

$$\delta W = \varphi\, \delta\rho$$

Physicists like to call this entity "work," but we will stick to using a less stringent definition of energy. We know from the boundary conditions, the charge on the conductor is

$$\rho = -\oint D_n\, dS = -\oint \mathbf{D}\cdot d\mathbf{S}$$

Here dS is a surface element and $d\mathbf{S}$ is surface element in the direction normal to the conductor surface. Since the potential is constant on the surface of the conductor we have

$$\delta W = \varphi\, \delta\rho = -\int \varphi\, \delta\mathbf{D}\cdot d\mathbf{S} = -\int \nabla\cdot(\varphi\, \delta\mathbf{D})dV$$

where in the last step we used Gauss' law and the integral extends over all volume. Outside the conductor Maxwell's equation states $\nabla\cdot\delta\mathbf{D} = 0$. By expanding the integrand in the equation we see

$$\nabla\cdot(\varphi\, \delta\mathbf{D}) = \varphi\nabla\cdot(\delta\mathbf{D}) + \delta\mathbf{D}\cdot\nabla\varphi = -\delta\mathbf{D}\cdot\mathbf{E}$$

We find

$$\delta W = \int \delta\mathbf{D}\cdot\mathbf{E}\, dV$$

We can now substitute $\boldsymbol{D} = \epsilon \boldsymbol{E}$ and by looking at the integrand above

$$\delta \boldsymbol{D}\cdot\boldsymbol{E} = \delta\epsilon \boldsymbol{E}\cdot\boldsymbol{E} = \epsilon\,\delta\boldsymbol{E}\cdot\boldsymbol{E} = \epsilon\frac{1}{2}\delta\,(\boldsymbol{E}\cdot\boldsymbol{E}) = \epsilon\frac{1}{2}\delta E^2$$

We have for the change in energy in the electric field

$$\delta W = \delta \int \epsilon \frac{1}{2} E^2 \, dV$$

We can so make plausible the definition of energy contained in an electric field to be

$$W_E = \frac{1}{2}\int \epsilon\, \boldsymbol{E}\cdot\boldsymbol{E}\, dV$$

Magnetic Field Energy

For a static magnetic field the situation is analogous. Here the work, or more roughly energy, done on currents is done by an electric field. A magnetic field by itself does no work on charges or current since it turns out that the magnetic force acting on a charge is perpendicular to its velocity. Instead we have to look for a situation where the magnetic field is varying with time, a quasi-static situation, and thus giving rise to an electric field through equation (4.3). We have in a time δt the energy spent by the external supplies

$$\delta W = -\delta t \int \boldsymbol{J}\cdot\boldsymbol{E}\, dV = -\delta t \int \nabla \times \boldsymbol{H}\cdot\boldsymbol{E}\, dV = \delta t \int \nabla \cdot (\boldsymbol{E} \times \boldsymbol{H})\, dV - \delta t \int \boldsymbol{H}\cdot(\nabla \times \boldsymbol{E})\, dV$$

The first term can be transformed into a surface integral at infinity which we assume is zero, all fields vanish there. The second term we transform

$$\delta W = -\delta t \int \boldsymbol{H}\cdot(\nabla \times \boldsymbol{E})\, dV = \int \boldsymbol{H}\cdot\delta \boldsymbol{B}\, dV$$

As we did for the electric field, we again integrate the integrand

$$\boldsymbol{H}\cdot\delta \boldsymbol{B} = \mu\frac{1}{2}\delta H^2$$

We get

$$\delta W = \delta \int \mu \frac{1}{2} H^2 dV$$

We can now analogously to the electric field energy make plausible the definition of energy in a magnetic field

$$W_M = \frac{1}{2}\int \mu\, \boldsymbol{H}\cdot\boldsymbol{H}\, dV$$

We now have

$$W_E = \frac{1}{2}\int \epsilon\, \boldsymbol{E}\cdot\boldsymbol{E}\, dV, \quad W_M = \frac{1}{2}\int \mu\, \boldsymbol{H}\cdot\boldsymbol{H}\, dV. \tag{4.42}$$

4.2 Maxwell's Equations

The integral extends over the total volume. This is the time-dependent definition of the energies. In microwave theory the *time average* definition is more useful and we can simply get there by taking the time average assuming a sinusoidal wave. We get another factor of ½ for a total of ¼ times the volume integrals.

Capacitance: Definition
Let us assume we have a simple configuration of two conductors with a voltage φ between them. We now define capacitance, C, as

$$\frac{1}{2} C \varphi^2 = W_E = \frac{1}{2} \int \epsilon \, \boldsymbol{E} \cdot \boldsymbol{E} \, dV \tag{4.43}$$

In the general case we have

$$\frac{1}{2} \sum_a \sum_b C_{ab} \varphi_{ab}^2 = \frac{1}{2} \int \epsilon \, \boldsymbol{E} \cdot \boldsymbol{E} \, dV = -\frac{1}{2} \int \epsilon \, \boldsymbol{E} \cdot \nabla \varphi \, dV = -\frac{1}{2} \int \nabla \cdot (\epsilon \boldsymbol{E} \varphi) dV$$

$$+ \frac{1}{2} \int \varphi \nabla \cdot (\epsilon \boldsymbol{E}) dV = \frac{1}{2} \int \varphi \rho \, dV = \frac{1}{2} \sum_a \sum_b \varphi_{ab} \rho_a$$

where in the last steps we used a common vector identity trick, $\nabla \cdot (\varphi \boldsymbol{b}) = \boldsymbol{b} \cdot \nabla \varphi + \varphi \nabla \cdot \boldsymbol{b}$. It has the convenient property that we only need to integrate over materials containing charges which greatly simplifies many calculations. By identifying terms we see capacitance is simply a linear relationship between charge and voltage. We will study this in more detail in the next section.

Key Concept

We need to know or estimate the voltages associated with various conductors and how the charge is distributed in order to estimate the capacitance.

Inductance: Definition
Analogously, for a simple situation with a wire carrying a current I, the inductance L can be defined as

$$\frac{1}{2} L J^2 = \frac{1}{2} \int \mu \, \boldsymbol{H} \cdot \boldsymbol{H} \, dV \tag{4.44}$$

For multiple currents J_a we instead have the general formula

$$\frac{1}{2} \sum_a \sum_b L_{ab} J_a J_b = \frac{1}{2} \int \mu \, \boldsymbol{H} \cdot \boldsymbol{H} \, dV = \frac{1}{2} \int (\nabla \times \boldsymbol{A}) \cdot \boldsymbol{H} \, dV$$

$$= -\frac{1}{2} \int \nabla \cdot (\boldsymbol{H} \times \boldsymbol{A}) dV + \frac{1}{2} \int \boldsymbol{A} \cdot (\nabla \times \boldsymbol{H}) dV = \frac{1}{2} \int \boldsymbol{A} \cdot \boldsymbol{J} \, dV$$

In previous steps we used a common trick to get rid of the large volume integrations by utilizing the vector identity: $\nabla \cdot (\boldsymbol{a} \times \boldsymbol{b}) = \boldsymbol{b} \cdot \nabla \times \boldsymbol{a} - \boldsymbol{a} \cdot \nabla \times \boldsymbol{b}$, and the divergence theorem coupled with the assumption the fields vanish at infinity. This shows we only need to integrate over materials where there is a current running. As we will see, this will greatly simplify some calculations.

If we calculate the contribution to A from each current separately we can write

$$A(x) = \sum_a A_a(x) \quad J(x) = \sum_b J_b(x)$$

And we find

$$\frac{1}{2}\sum_a \sum_b L_{ab} J_a J_b = \frac{1}{2}\sum_a \sum_b \int A_a \cdot J_b \, dV \qquad (4.45)$$

where if $a \neq b$ we calculate the mutual inductance between elements a, b and if $a = b$ we speak of self-inductance of element a. We will use these facts later in this section.

$$L_{self,a} \; J_a J_a = \int A_a \cdot J_a \, dV \quad L_{mutual,ab} \; J_a J_b = \int A_a \cdot J_b \, dV \qquad (4.46)$$

Please note that when the vector potential from a certain conductor, say a, is perpendicular to the current in another conductor b the contribution from that term in (4.45) is zero. This is also an important simplification in many cases.

$$\int A_a \cdot J_b \, dV \equiv 0 \quad \text{when} \quad A_a \perp J_b$$

What the reader should take home from these calculations is the physical analogy between capacitance and inductance. One is based on voltage and charge the other on currents and vector potential. If you understand one of them you are likely to understand the other.

Key Concept

We know from elementary classes that currents will follow the least impedance path. In this context this simple rule means that **the current will flow in way that minimizes inductance** if there are no other effects like resistance/capacitance to consider.

We need to know or estimate how the currents are flowing in order to estimate the inductance.

The way to estimate inductance now becomes a way to estimate how the currents are flowing in the model. We will look at a number of situations where we will most of the time start directly from Maxwell's equations and from there learn how the current distributes itself.

4.3 Capacitance

We have now finished with basic definitions of Maxwell's equations and their solution and defined the concepts of capacitance and inductance. In the rest of the chapter we will apply the estimation analysis technique to explore these concepts more practically. We will find a way to simplify a given situation and we will solve and verify the solution. This will be followed by an evaluation phase. First we tackle something most likely very familiar to most engineers: capacitance.

Introduction

Most working engineers have a good feel what capacitance is and we will discuss it in ways that are very familiar with most readers and also use approaches that might not be as familiar but will prove to be good starting points and learning experiences. First we describe the circuit element: the capacitor. We then discuss a simple two plate system with a voltage between them which is an expansion of the discussion in section "Field Energy Definitions." Thereafter we will solve the same problem using Maxwell's equations directly where we also use the boundary conditions developed in section "Solutions to Maxwell's Equations." We will finally show how the same approach can be used when investigating an example with two different dielectrics.

Capacitors as a Circuit Element

So far we have discussed capacitance as it relates to physics: it is an effect that draws charge to a conductive surface when a voltage is applied to some other conductor. But what about capacitance as a circuit element? To investigate this we will use a simple conservation of power (or energy) argument. Imagine we have a resistor in parallel with a capacitor we know the power spent in the resistor is

$$P_R = \frac{\varphi^2}{R}$$

where φ is the voltage across the resistor and capacitor. The power in the capacitor is simply the time derivative of its energy

$$P_C = \frac{d}{dt} \frac{1}{2} C \varphi(t)^2 = C\varphi \frac{d\varphi}{dt}$$

The voltage source that drives this combination is having a power $\varphi(t) * I(t)$ pulled from it at a given time. This power has to equal the power spent in the resistor and capacitor:

$$\varphi(t)I(t) = \frac{\varphi^2}{R} + C\varphi \frac{d\varphi}{dt}$$

or

$$I(t) = \frac{\varphi(t)}{R} + C\frac{d\varphi(t)}{dt}$$

This is the familiar circuit relationship. It becomes clearer if one assumes a time dependence $e^{j\omega t}$.

$$I = \frac{\varphi}{R} + j\omega C\, \varphi$$

where the impedance due to capacitance is

$$Z_c = \frac{\varphi}{I} = \frac{1}{j\omega C}$$

We have shown the circuit formula for using capacitors follows simply from a conservation of energy argument.

Simple Two-Plate System Calculation

Let us use as a starting point equation (4.43). To get a feel for what these relationships mean in a real situation let us consider two metal plates of area A, a distance, d, from each other with a voltage φ between them. The electric field is $E = \varphi/d$. We then have

$$\frac{1}{2} C\, \varphi^2 = W_E = \frac{1}{2}\int \epsilon \left(\frac{\varphi}{d}\right)^2 dV \approx \frac{1}{2}\epsilon \left(\frac{\varphi}{d}\right)^2 A d = \frac{1}{2}\epsilon \frac{\varphi^2}{d} A$$

or

$$C = \frac{\epsilon}{d} A. \tag{4.47}$$

This is the normal calculation of capacitance known from elementary classes.

First Principle Calculation of Capacitance of Two-Plate System

Let us know look at the same situation from first principles, in this case Maxwell's equations. We will follow the estimation analysis method. It will seem a little excessive at first, in particular when comparing with the previous calculation, but we will show we can easily extend what we learn here to other situations with little extra effort.

Simplify The first thing to do is simplification. Let us imagine the two plates are infinitely extended in all directions in the plane. That will simplify the problem to one dimension. Next the top plate has a voltage, V while the bottom plate is grounded. Lastly we assume the long wavelength approximation so we need to solve equation (4.26). We find as in Figure 4.2.

In free space assuming no x-dependency we have then

$$\Delta \varphi = -\frac{C\, \delta(y - y_0)}{\epsilon'}$$

Subject to the boundary conditions

$$\varphi(y_0) = V \quad \varphi(0) = 0$$

Solve This equation is independent of z and is a one-dimensional Helmholz equation in the long wavelength approximation which makes it Poisson's equation and it has solution:

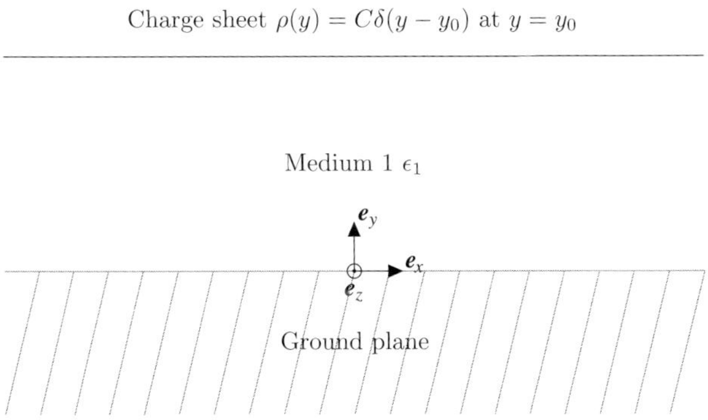

4.3 Capacitance

Charge sheet $\rho(y) = C\delta(y - y_0)$ at $y = y_0$

Medium 1 ϵ_1

Ground plane

Figure 4.2 Two-dimensional projection of two plates.

$$\varphi(y) = -A(y - y_0) + B, \quad \text{for } y_0 \geq y$$

Let us plug in the boundary conditions

$$\begin{cases} B = V \\ Ay_0 + B = 0 \end{cases}$$

$$A = -\frac{V}{y_0}$$

We get for

$$\varphi(y) = -A(y - y_0) + B$$

The electric field is $E = -\nabla\varphi$. At the lower perfect electrical conductor (PEC) boundary the electric field will change abruptly from the gradient of the potential to zero. This will result in surface charge and it follows from the boundary condition on $\varepsilon\nabla\cdot E = \rho$ which gives $\epsilon E_+ = \rho$. In effect what is happening is the voltage on the top is inducing a charge in the bottom plate. At long wavelength this is simply known as a capacitor effect. We will go through the simple derivation now.

We have

$$|E| = |-\nabla\varphi| = A = -\frac{V}{y_0}.$$

Since we now know the electric field we can simply plug it into the formula for capacitance (4.43) and recover equation (4.47). Going further we can use our more sophisticated model to recover another formula involving capacitance. Let us put a pill box around the boundary and integrating we find:

$$\varepsilon \int \nabla\cdot E \, dV = \varepsilon\left(E\big|_{y=0+} - E\big|_{y=0-}\right) Area = \varepsilon \, E\big|_{y=0+} Area$$

$$= -\varepsilon\frac{V}{y_0} Area = \int \rho \, dV = \int \rho_0 \delta(y=0) dV = Q.$$

We have

$$-\varepsilon \frac{V}{y_0} Area = Q$$

where $Q < 0$. If we identify $\frac{\varepsilon}{y_0} Area = C$ or capacitance we have

$$CV = Q. \qquad (4.48)$$

This is another famous relationship involving capacitance and we see how straightforward it is to derive it directly from Maxwell's equations.

Verify This is the same result we saw earlier, leading to equation (4.47), but this was a more complicated calculation since we started from scratch. The advantage is we have learned a way to use Maxwell's equations directly and we are now equipped to solve more complex problems.

Evaluate The capacitance between two metal plates scales as the overlapping area of the plates divided by their distance.

First Principle Calculation of Capacitance with Two Different Dielectric Media

Here we will examine how the field solutions behave when there are different dielectric media in the problem. This situation often shows up in integrated circuits where there are different dielectric layers and instead of implementing all of them in a field solver it is often enough to use the equivalent permittivity. Here we will show how to calculate the effective permittivity and we will again start from Maxwell's equations and end up with an expression that is quite familiar to most readers.

Simplify We build on the previous simple model ad simply add another boundary: see Figure 4.3. All will still be one-dimensional.

Compared with previous calculations, everything is the same – the same equations, the same solutions – but we have an additional boundary condition at the interface between the two dielectrics. From $\nabla \cdot \varepsilon E = \rho$ we find

$$\varepsilon_1 E|_1 = \varepsilon_2 E|_2$$

since there is no surface charge in the dielectric media.

Solve We now have two region and two solutions:

$$\varphi_1(y) = -A_1(y - y_0) + B_1$$
$$\varphi_1(y) = -A_2 y + B_2$$

For boundary conditions we get

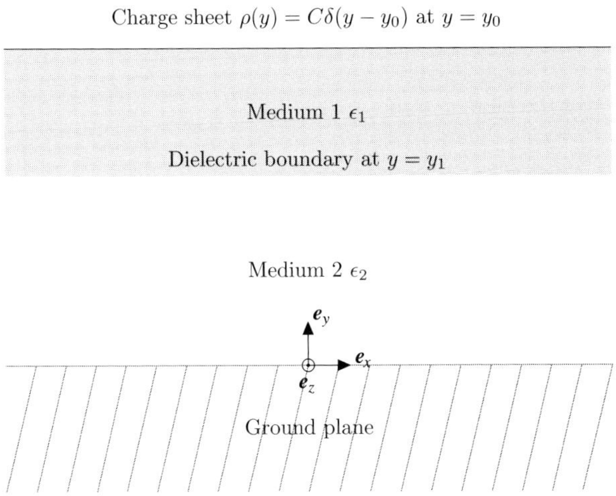

4.3 Capacitance

Charge sheet $\rho(y) = C\delta(y - y_0)$ at $y = y_0$

Medium 1 ϵ_1

Dielectric boundary at $y = y_1$

Medium 2 ϵ_2

Ground plane

Figure 4.3 Two-dimensional picture of two plates with two dielectric media.

$$\begin{cases} B_1 = V \\ B_2 = 0 \\ \varepsilon_1 E|_1 = \varepsilon_2 E|_2 \\ \varphi_1(y_1) = \varphi_2(y_1) \end{cases}$$

The third one can be expanded:

$$\varepsilon_1 E|_1 = -\varepsilon_1 \nabla \varphi_1 = \varepsilon_1 A_1 = \varepsilon_2 E|_2 = \varepsilon_2 A_2$$

or

$$A_2 = \frac{\varepsilon_1}{\varepsilon_2} A_1$$

We find from the fourth boundary condition

$$-A_1(y_1 - y_0) + V = -\frac{\varepsilon_1}{\varepsilon_2} A_1 y_1$$

which gives

$$A_1 = -\frac{V}{\left(y_0 - y_1 + \frac{\varepsilon_1}{\varepsilon_2} y_1\right)}$$

and

$$A_2 = -\frac{\varepsilon_1}{\varepsilon_2} \frac{V}{\left(y_0 - y_1 + \frac{\varepsilon_1}{\varepsilon_2} y_1\right)}$$

We find for the electric field at $y = 0$

$$E = -\nabla \varphi_2 = -\frac{\varepsilon_1 V}{y_1 \varepsilon_1 + \varepsilon_2 (y_0 - y_1)}$$

As before we have for the charge at the lower PEC

$$\varepsilon_2 E \, Area = -\frac{\varepsilon_1 \varepsilon_2}{y_1 \varepsilon_1 + \varepsilon_2 (y_0 - y_1)} V Area = Q$$

We can rewrite this in more familiar form:

$$-\frac{1}{y_1/Area \, \varepsilon_2 + (y_0 - y_1)/Area \, \varepsilon_1} V = -\frac{1}{1/C_1 + 1/C_2} V = Q$$

This is the well-known serial formula for capacitors. We can furthermore rewrite the equation in a form that makes use of an effective permittivity, ε'

$$\frac{\varepsilon' V}{y_0} Area = \frac{\varepsilon_1 \varepsilon_2}{y_1 \varepsilon_1 + \varepsilon_2 (y_0 - y_1)} V Area$$

which gives

$$\varepsilon' = \frac{\varepsilon_1 \varepsilon_2}{y_1 \varepsilon_1 + \varepsilon_2 (y_0 - y_1)} y_0 = \frac{1}{y_1/\varepsilon_2 + (y_0 - y_1)/\varepsilon_1} y_0.$$

Verify This is simply a rederivation of the well-known result of the equivalent capacitance of two capacitors in series. It is demonstrated in most elementary textbooks on electronics.

Evaluate When stacking dielectrics with different permittivity the effective permittivity can be calculated as the inverse of a weighted sum of inverses.

Key Concept

When stacking dielectrics with different permittivity the effective permittivity can be calculated as the inverse of a weighted sum of inverses.

Summary
We have applied the estimation analysis to two situations where the total capacitance was needed. We calculated it directly from Maxwell's equations using some basic simplifications as directed by the estimation analysis technique.

4.4 Inductance

We continue here with the concept of inductance. We have defined it in section "Field Energy Definitions" and we will here explore the definitions further looking

4.4 Inductance

at some simple situations where our estimation analysis technique will prove helpful.

Introduction

This section discusses the concept of inductance. As this concept may not be familiar to most engineers we hope to demystify it here. We will follow the outline in Section 4.3 to further show the analogy between inductance and capacitance, where we first discuss the circuit element – inductor – and then describe the simplest model followed by a treatment starting from Maxwell's equations directly. Equipped with this we can attack more complicated problems. Much of the details here can be found in [4]. The inductance effect can be very detrimental to high-speed circuit behavior and one must understand its nature and how it can become large in order to produce successful designs.

Inductors as Circuit Elements

So far we have discussed inductors as it relates to physics: it is an effect that is proportional to the magnetic energy from a current distribution. It is not immediately obvious how it is related to circuit analysis. To investigate this we will use a simple conservation of power (or energy) argument.

Imagine we have a resistor in series with an inductor we know the power spent in the resistor is

$$P_R = I^2 R$$

where I is the current across the resistor and capacitor. The power in the inductor is simply the time derivative of its energy

$$P_C = \frac{d}{dt}\frac{1}{2}LI^2 = LI\frac{dI}{dt}$$

The voltage source that drives this combination is having a power $\varphi(t)*I(t)$ pulled from it at a given time. This power has to equal the power spent in the resistor and capacitor:

$$\varphi(t)I(t) = I^2 R + LI\frac{dI}{dt}$$

or

$$\varphi(t) = IR + L\frac{dI(t)}{dt}.$$

This is the familiar circuit relationship. It becomes clearer if one assume a time dependence $e^{j\omega t}$.

$$\varphi(t) = IR + j\omega LI$$

Where the impedance due to inductance is

$$Z_c = \frac{\varphi}{I} = j\omega L.$$

We have shown the circuit formula for using inductors follows simply from a conservation of energy argument.

Simple Straight Wire in Free Space

We studied capacitance earlier in this chapter. Inductance is only slightly different. Instead of an electric field and charge, inductance relates to magnetic field and current. An electric field will look for other potentials on conductors around it to terminate, a magnetic field terminates on itself. This self-termination, which we will explore details below, can cause the magnetic field to become unbound. This has dramatic impact on the generating currents and how they can be allowed to behave.

Before we get into the mathematical details let us play a thought experiment. Imagine there is a universe with one unique feature: there is nothing in this universe but a wire, carrying a current, I. The wire is infinite and straight. We will assume Maxwell's equations are valid. What is going on? Well, there is a current, so there must be a magnetic field (one of Maxwell's laws). The converse is also true, if there is a magnetic field there has to be a current supporting it. The field stretches forever and ever around this wire. The total magnetic field energy,

$$W_M = \frac{1}{2}\int \mu \mathbf{H}\cdot\mathbf{H}\, dV = \left\{ \mathbf{H} = \frac{I}{2\pi r}\, \mathbf{e}_\theta\ (\text{We will derive this later})\right\}$$
$$= \frac{1}{2}\int \mu \left(\frac{I}{2\pi r}\right)^2 r\, dr d\theta \sim \ln R \quad \text{where } R \text{ is the size of the wire} \to \infty \text{ in our case.}$$

What does this mean? If the magnetic field energy is infinite, so is the inductance. If you attempt to drive an AC current through this wire you will see an infinite impedance. In our simple universe, nothing can ever happen!

Admittedly this is far-fetched, but it serves to illustrate a point. In order for wires and currents to be useful there has to be a way to limit the far-field magnetic field. The way this is usually done is through loops either a current loop in itself or an induced return path in some ground plane. There has to be a closed current loop for circuitry to work! This is unlikely a surprise.

First Principle Calculation of a Simple Straight Wire

We will now dig deeper into the details of the previous section by starting from Maxwell's equations directly using our estimation analysis.

Simplify We have already set up a situation that is relatively simple. A straight wire in free space with uniform current distribution. We will model it as a system with no z-dependence, a two-dimensional system with cylindrical symmetry so all entities only depend on the distance, r, from the center of the wire. Seen from the side we find as in Figure 4.4.

Let us look at this simple wire and dig into the mathematical details. We only need Ampere's law in the long wavelength approximation

$$\nabla \times \boldsymbol{H} = \boldsymbol{J}$$

Solve We can integrate Ampere's law over a cross-section containing the wire as in Figure 4.5:

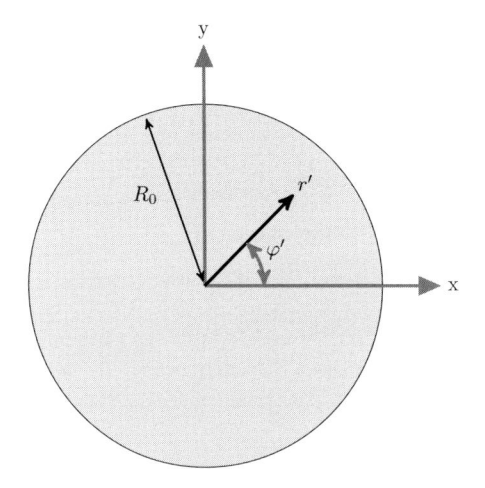

Figure 4.4 Cross-section of single wire.

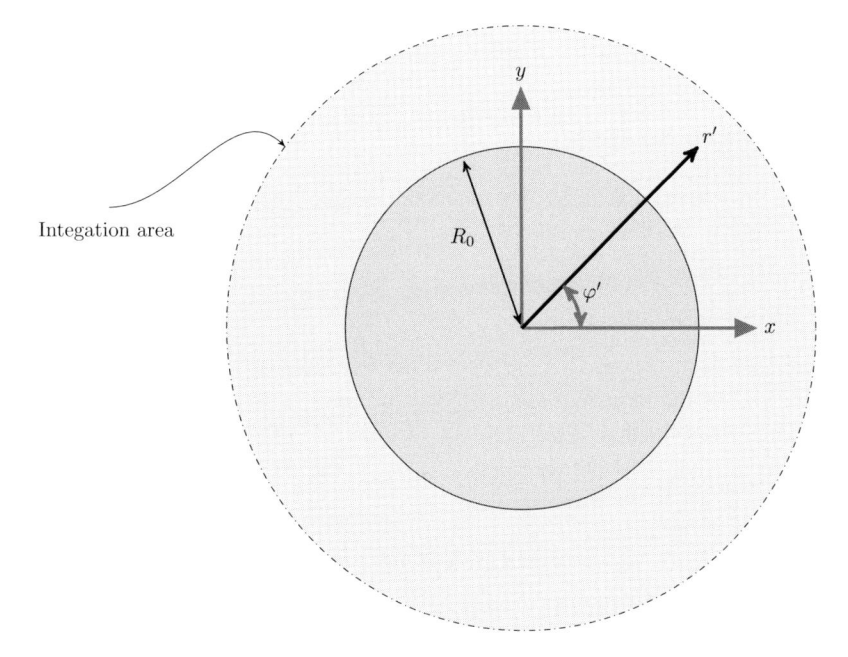

Figure 4.5 Integration boundary of single wire.

$$\int \nabla \times \boldsymbol{H} \cdot d\boldsymbol{S} = \int \boldsymbol{J} \cdot d\boldsymbol{S}$$

We can use Stoke's theorem on the curl side:

$$\int \nabla \times \boldsymbol{H} \cdot d\boldsymbol{A} = \oint \boldsymbol{H} \cdot d\boldsymbol{r} = \oint H_\varphi(r)\, r\, d\varphi = 2\pi r H_\varphi(r)$$

where we have taken advantage of the cylindrical symmetry. On the right-hand side we get from the fact the current, $\boldsymbol{J}$, is a density

$$\int \boldsymbol{J} \cdot d\boldsymbol{S} = J_{total}$$

Putting the two sides together we find

$$H_\varphi(r) = \frac{J_{total}}{2\pi\, r}$$

It is interesting to see here this result is independent of the precise radial distribution of the current internal to the conductor. The right-hand side is only sensitive to the total current in conductor.

Key Concept

The external magnetic field is not dependent on the radial current distribution inside the conductor in the cylindrically symmetric case.

This in turn implies that the inductance from the total magnetic energy is.

$$\frac{1}{2}LJ_{total}^2 = \frac{1}{2}\int \mu \boldsymbol{H}\cdot\boldsymbol{H}\, dV = \frac{1}{2}\mu l \left(\frac{1}{2\pi}\right)^2 \int \left(\frac{J_{total}}{r}\right)^2 r\, dr d\theta = \frac{1}{2}\frac{1}{2\pi}\mu l J_{total}^2 \ln(R).$$

Here we have excluded the magnetic energy internal to the wire. It makes little difference. We find

$$L = \frac{\mu l}{2\pi}\ln(R)$$

where R is the "extent" of the wire. It has to be a closed loop somehow and R is simply the size of the loop. We see the inductance will be infinite for a single wire without a loop as we hinted at in the previous section.

As a quick side note we can also calculate the vector potential outside the conductor when we know the magnetic field

$$\nabla \times \boldsymbol{A} = \boldsymbol{B} = \mu \boldsymbol{H} \quad \rightarrow$$
$$\boldsymbol{A} = A(r)\boldsymbol{e}_z = \left(-\mu \frac{J_{total}}{2\pi}\ln\frac{r}{R_0} + C\right)\boldsymbol{e}_z. \tag{4.49}$$

Verify This is a known solution and can be found in for example [4].

4.4 Inductance

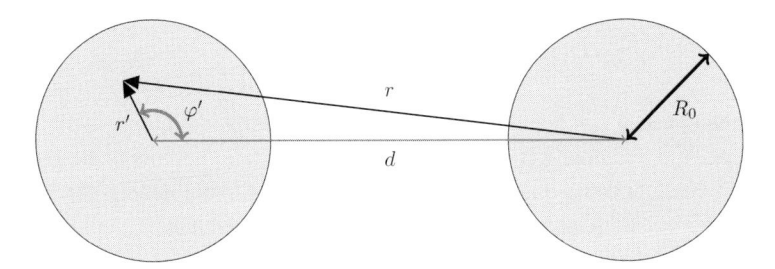

Figure 4.6 Cross-section of two wires in two dimensions. Figure adapted from [16].

Evaluate The inductance for a wire scales as the length of the wire times the logarithm of its size.

First Principle Calculation of Two Simple Straight Wires

Imagine now we have two wires where the current in one goes the opposite direction as the current in the other. The idea is to reduce the magnetic field in the single wire by creating an opposing field that is almost aligned. In effect we have a local current loop.

Simplify We use the same basic model as previously but with the addition of one wire. The internal current distribution is assumed to be uniform: see Figure 4.6.

Solve To calculate the inductance of the system we will use equation (4.46) we derived earlier. The equation states that knowing the vector potential and the current distribution inside the conductors we can calculate the inductance through a simple integration over the conductors cross-sections only. We have already assumed the current distribution is uniform and we are left with calculating the vector potential. The vector potential external to the wires we already know from equation (4.49). Let us also calculate A inside a conductor. We have using the same strategy as in the previous section where we integrated Ampere's law and used Stoke's theorem:

$$2\pi H(r) r = \int \mathbf{J} \cdot d\mathbf{S} = \frac{J_{total}}{\pi R_0^2} \int_0^{R_0} r^2 2\pi r' dr' = \frac{J_{total}}{R_0^2}$$

where we have used the current density

$$|\mathbf{J}| = \frac{J_{total}}{\pi R_0^2}$$

and $r \leq R_0$.
We get

$$H(r) = \frac{J_{total}}{2\pi R_0^2} r.$$

From $\mathbf{B} = \mu \mathbf{H} = \nabla \times \mathbf{A}$ we find

$$\mu \frac{J_{total}}{2\pi R_0^2} r = -\frac{\partial A}{\partial r} \quad \rightarrow \quad A = C + \mu \frac{J_{total}}{2\pi} \frac{1}{2} - \mu \frac{J_{total}}{2\pi R_0^2} \frac{r^2}{2}$$

where the constant has been chosen to match the external solution (4.49) at the $r = R_0$ boundary. We can now put things together to calculate the energy by integrating the self-inductance and the mutual inductance from (4.46). We denote the energy for the self-inductance, F_{self}, and the energy for the mutual inductance calculation, F_{mutual}.

$$F_{self} = \frac{1}{2} \int_0^{R_0} \left(C + \mu \frac{J_{total}}{2\pi} \frac{1}{2} - \mu \frac{J_{total}}{2\pi R_0^2} \frac{r'^2}{2} \right) \frac{J_{total}}{\pi R_0^2} r' dr' 2\pi Z$$

$$F_{mutual} = -\frac{1}{2} \int_0^{2\pi} \int_0^{R_0} \left(-\frac{\mu J_{total}}{2\pi} \ln(r/R_0) + C \right) \frac{J_{total}}{\pi R_0^2} r' dr' d\varphi' Z$$

where Z is the unit length of the wire. The variable r is the distance between the center of conductor 2 and r', φ' as indicated in the figure.

$$r^2 = d^2 + r'^2 - 2dr' \cos \varphi'$$

$$F_{self} = \frac{1}{2} \frac{J_{total}}{\pi R_0^2} 2\pi Z \int_0^{R_0} \left(C + \mu \frac{J_{total}}{2\pi} \frac{1}{2} - \mu \frac{J_{total}}{2\pi R_0^2} \frac{r'^2}{2} \right) r' dr' \qquad (4.50)$$

$$= J_{total} ZC \frac{1}{2} + \mu \frac{J_{total}^2}{2\pi} Z \frac{1}{8}$$

$$F_{mutual} = -\frac{1}{2} \frac{J_{total}}{\pi R_0^2} 2\pi ZC \frac{R_0^2}{2} + \mu \frac{1}{2} \frac{J_{total}^2}{2\pi} \frac{1}{\pi R_0^2} Z \int_0^{2\pi} \int_0^{R_0} \frac{1}{2} \ln\left(\frac{d^2 + r'^2 - 2dr' \cos \varphi'}{R_0^2} \right) r' dr' d\varphi'.$$

The integral in the expression for F_{mutual} can be calculated by identifying

$$I(d) = \int_0^{2\pi} \int_0^{R_0} \frac{1}{2} \ln\left(\frac{d^2 + r'^2 - 2dr' \cos \varphi'}{R_0^2} \right) r' dr' d\varphi'.$$

This is straightforward to calculate and we find

$$I(d) = \pi R_0^2 \ln \frac{d}{R_0}.$$

Now

$$F_{mutual} = -\frac{1}{2} \frac{J_{total}}{\pi R_0^2} 2\pi ZC \frac{R_0^2}{2} + \mu \frac{1}{2} \frac{J_{total}^2}{2\pi} \frac{1}{\pi R_0^2} Z \pi R_0^2 \ln \frac{d}{R_0} \qquad (4.51)$$

$$= -J_{total} ZC \frac{1}{2} + \mu \frac{1}{4\pi} J_{total}^2 Z \ln \frac{d}{R_0}.$$

The total energy is now

$$F = F_{self} + F_{mutual} = \mu \frac{1}{2\pi 2} J_{total}^2 Z \left(\frac{1}{4} + \ln\left(\frac{d}{R_0}\right) \right).$$

The other conductor combination is now easy to include because of symmetry, we just multiply by 2. By removing $J_{total}^2/2$ we find the inductance is

$$L = \frac{\mu Z}{\pi} \left(\frac{1}{4} + \ln\left(\frac{d}{R_0}\right) \right) \tag{4.52}$$

This is one of our fundamental results and we will use this again and again in various contexts.

Verify This is a standard calculation in physics literature: see [4, 15], for example. For the reader coming from a microwave background this example might seem familiar. If so, it should, but the microwave literature often assumes the conductors to be ideal and in that case the solution is somewhat different. We leave it as an exercise for the reader to solve the ideal conductor case.

Evaluate For two inductors on top of each other, if the current is going the same way in both inductors the magnetic field is doubled (with equal current) → magnetic energy is quadrupled → inductance is four times larger. If currents are going the opposite way → magnetic field is nulled → inductance is zero.

This effect is often referred to as coupling between the wires. It comes from the high frequency case where currents are induced in (or coupled to) the neighboring conductors.

Key Concept

For two inductors on top of each other, if the current is going the same way in both inductors the magnetic field is doubled (with equal current) → magnetic energy is quadrupled → inductance is four times larger. If currents are going the opposite way → magnetic field is nulled → inductance is zero.

This effect is often referred to as coupling between the wires. It comes from the high frequency case where currents are induced in (or coupled to) in the neighboring conductors.

We will study the high frequency case and induced currents later in this chapter.

First Principle Calculation of Single Wire over Ground Plane

When a wire is running over a ground plane, we have a very similar situation to the previous discussion: see Figure 4.7.

Simplify Using the method of images we can model the situation in exactly the same way where the distance d is simply

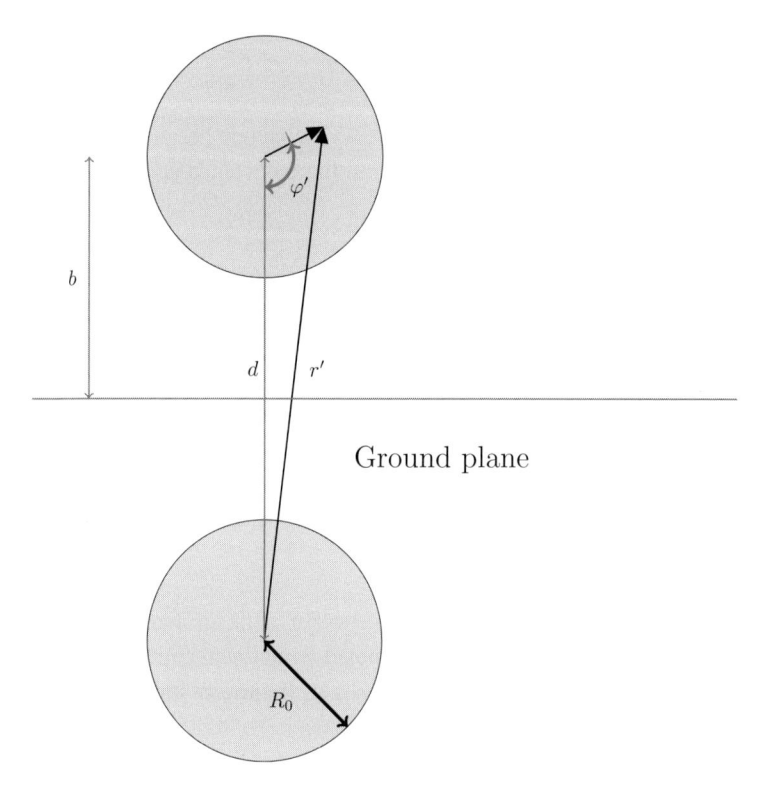

Figure 4.7 Method of images demonstrated with one wire over a ground plane.

$$d = 2\,b$$

where b as in the figure is the distance to the ground plane from the wire. The method of images, [1], simply states the field *outside a perfect ground plane* can be found by removing the ground plane and put mirror conductors with opposite charge/current equidistant from the ground plane border.

Solve We can now use exactly the same calculations as before with one important difference. The total field energy is now half of what we had before since no field exists in the ground plane which takes up half the volume.

We find

$$L_{ground} = \frac{1}{2}\frac{\mu Z}{\pi}\left(\frac{1}{4} + \ln\left(\frac{d}{R_0}\right)\right). \tag{4.53}$$

Verify This situation has been simulated as a $0.1 \rightarrow 4$ mm long wire (1 μm × 1 μm cross-section) over a ground plane in HFSS: see Figure 4.8.

The conductor was a variable height over the ground plane. The excitation is through wave ports at the end of the structure as indicated. The inductance is calculated as

4.4 Inductance

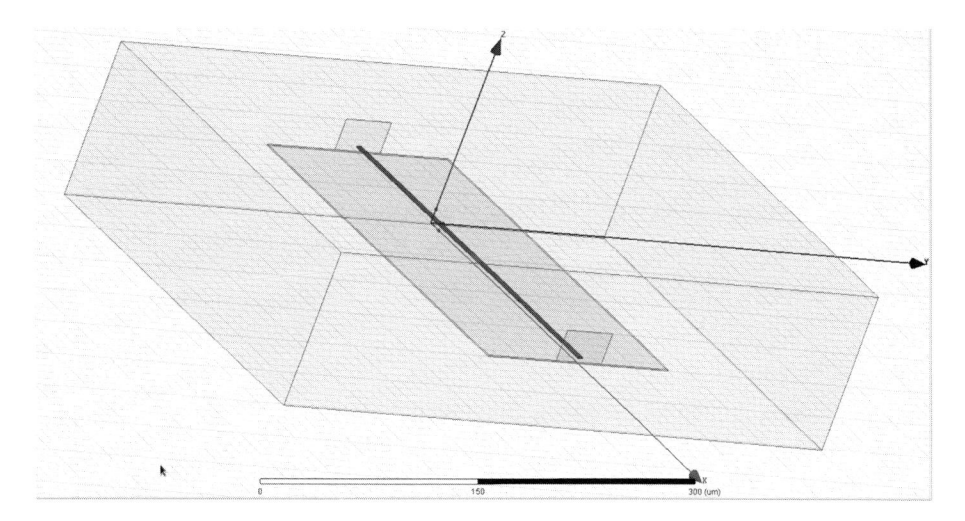

Figure 4.8 Figure of an HFSS sim setup.

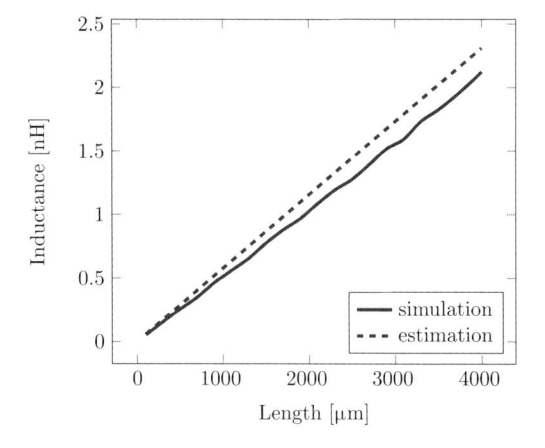

Figure 4.9 Simulated and estimated inductance of single wire over ground plane vs length.

$$L_{sim} = \text{im}\left(\frac{-2}{Y(1,2) + Y(2,1)}\right)\frac{1}{\omega}$$

where Y refers to the Y-parameter (admittance) gain. Figure 4.9 shows the simulation comparison to (4.53) as a function of length where the height over the ground plane is 3 μm.

A comparison of simulation vs estimation of inductance vs height over ground plane shows in Figure 4.10.

Evaluate When adding a ground plane under a conducting wire the resulting inductance will be reduced compared to a single wire. The inductance scales roughly as the logarithm of the distance to the ground plane. This is for the case of no other conductors nearby.

> **Key Concept**
>
> When adding a ground plane under a conducting wire the resulting inductance will be reduced compared to a single wire. The inductance scales roughly as the logarithm of the distance to the ground plane.

First Principle Calculation of a Current Sheet over a Ground Plane

Let us now look at the current analogue to a charge plane over a ground plane. This is the situation described in section "Simple Two-Plate System Calculation," but now with a current sheet instead of a charge sheet (Figure 4.11).

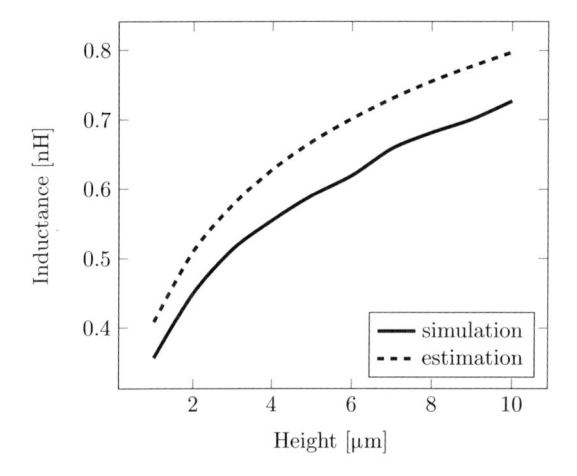

Figure 4.10 Inductance of single wire over ground plane vs height over ground plane.

Current sheet $i(y) = J\delta(y - y_0)$ at $y = y_0$

Medium 1, μ_1

Ground plane

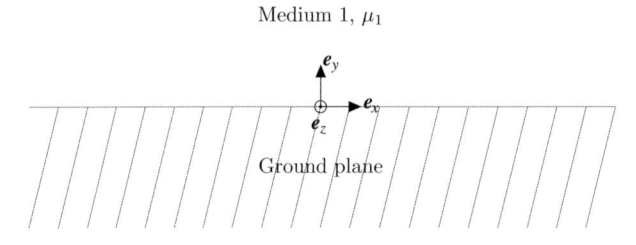

Figure 4.11 Cross-section of current sheet over ground plane.

4.4 Inductance

Simplify Let us know look at the basic equations in terms of the vector potential $\mathbf{A} = A(y)\mathbf{e}_z$, where the source term is a current. We could utilize the same simplifications we used in the capacitance calculation but here we will go one step further and start with the full wave equation and later go to the long wavelength limit so we can demonstrate some new mathematical steps.

In free space assuming no x-dependency we have

$$\Delta A + \omega^2 \epsilon' \mu A = -\mu J \delta(y - y_0)$$

We have

$$\frac{\partial^2}{\partial y^2} A + \omega^2 \epsilon' \mu A = -\mu J \delta(y - y_0)$$

Solve We can use

$$\omega^2 \epsilon' \mu = \kappa^2$$

$$\frac{\partial^2}{\partial y^2} A + \kappa^2 A = -\mu J \delta(y - y_0)$$

This equation has the solution from equation (4.31).

$$A(y) = \mu \frac{j}{2\kappa} e^{j\kappa |y-y_0|} + B e^{j\kappa y} = C e^{j\kappa |y-y_0|} + B e^{j\kappa y} = C e^{j\kappa (y_0-y)} + B e^{j\kappa y}, \quad \text{for } y_0 \geq y$$

For the boundary conditions we have

$$\nabla \times \mathbf{B} \to -B_x \big|_{y=y_0} = J_{y=y_0}/\text{Area}, \quad B_x \big|_{y=0} = J_{y=0}/\text{Area}, \quad J_{y=y_0} = -J_{y=y_0} = -J.$$

We get

$$B_x(y) = \frac{\partial A(y)}{\partial y} = -j\kappa C e^{j\kappa(y_0-y)} + j\kappa B e^{j\kappa y}.$$

From the boundary conditions we find:

$$\begin{cases} -j\kappa C + j\kappa B e^{j\kappa y_0} = \dfrac{J}{\text{Area}} \\ -j\kappa C e^{j\kappa y_0} + j\kappa B = \dfrac{J}{\text{Area}} \end{cases}$$

Solving for the constant B we get:

$$j\kappa B e^{j2\kappa y_0} - j\kappa B = \frac{J}{\text{Area}} \left(e^{j\kappa y_0} - 1 \right) \quad \to \quad B = \frac{J}{j\kappa \text{Area}} \frac{(e^{j\kappa y_0} - 1)}{e^{j2\kappa y_0} - 1}$$

and

$$j\kappa C = j\kappa B e^{j\kappa y_0} - \frac{J}{\text{Area}} \quad \to \quad C = \frac{J}{j\kappa \text{ Area}} \left(\frac{e^{j\kappa y_0} - 1}{e^{j2\kappa y_0} - 1} - 1 \right) = \frac{J}{j\kappa \text{ Area}} e^{j\kappa y_0} \left(\frac{1 - e^{j\kappa y_0}}{e^{j2\kappa y_0} - 1} \right).$$

We have

$$B_x(y) = \frac{\partial A(y)}{\partial y} = -\frac{J}{Area}e^{j\kappa y_0}\left(\frac{1-e^{j\kappa y_0}}{e^{j2\kappa y_0}-1}\right)e^{j\kappa(y_0-y)} + \frac{J}{Area}\frac{e^{j\kappa y_0}-1}{e^{j2\kappa y_0}-1}e^{j\kappa y}.$$

At the long wavelength approximation we can simplify

$$B_x(y) \approx -\frac{J}{Area}\left(\frac{-j\kappa y_0}{j2\kappa y_0}\right)(1+j\kappa(y_0-y)) + \frac{J}{Area}\,j\kappa y_0\frac{(1+j\kappa y)}{j2\kappa y_0}$$
$$= \frac{J}{Area}\frac{(2+j\kappa y_0)}{2} \approx \frac{J}{Area}.$$

By integrating the magnetic energy over the region we find:

$$\int_0^{y_0} B^2\, dV = \left(\frac{J}{Area}\right)^2 \int_0^{y_0} Area\, dy = \left(\frac{J}{Area}\right)^2 Area\, y_0 = \frac{J^2}{Area}y_0.$$

From the definition of inductance, equation (4.4) we find

$$L = \mu \frac{1}{Area} y_0. \tag{4.54}$$

The steps we followed here were more complicated than absolutely necessary to get to this answer. However, we have demonstrated a full solution to this version of Maxwell's equations, in this case Helmholz equation, and we will look more into this kind of calculation in the rest of this chapter.

Verify We can now see an interesting relationship with the calculation of capacitance between two plates. If we multiply equations (4.47) and (4.54):

$$LC = \mu\frac{1}{Area}y_0\,\frac{\epsilon}{y_0}Area = \mu\epsilon \tag{4.55}$$

Evaluate This turns out to be a general relationship for two dimensions, **any two shapes have an inductance per unit length times a capacitance per unit length that is equal to $\mu\epsilon$ in *two dimensions***. The proof of this goes beyond the scope of this book but can be found in many references. It is a very useful rule to keep in mind. In practice it also holds up well for planar geometries in three dimensions where one has long skinny conductors. This is an example of a situation where estimation analysis yields a result that turns out to be quite general. If you find some simple relationship like the one just descrbied try to see if it is more general than your simplifications imply. Perhaps you have discovered something fundamental?

> **Key Concept**
>
> Any two shapes have an inductance per unit length times a capacitance per unit length that is equal to $\mu\epsilon = \frac{1}{c^2}$ for two dimensions. In three dimensions it holds well for planar geometries with long skinny conductors.

Summary

We have applied the estimation analysis to a few situations where the total inductance was needed. We calculated it directly from Maxwell's equations using some basic simplifications as directed by the estimation analysis technique.

4.5 Various High Frequency Phenomena

Introduction

In this section we will study various high frequency phenomena that are of interest to the integrated circuit designer using estimation analysis. We will not consider frequencies so high the dielectric properties of the media changes. We will derive the skin depth phenomena from first principles in the first section. This will be followed by a study of induced current in a perfect ground plane, all from Maxwell's equations or use solutions thereof studied earlier in this chapter. We will also study the general case of full wavelength approximation and how a resistive ground plane distributes its return current in the following section. We will finally see how currents in a thin metal wire, relevant to modern CMOS metal stack-ups, gets distributed.

Skin Depth

We have so far used the static approximation when studying inductance and come up with some useful approximations and concepts that will help us get a grip on the concept. Most of the time the wavelengths involved in IC design are large compared with the chip dimensions. However, there is one important exception due to the finite conductance in the routing layers giving rise to the "skin" effect. We will study this concept from the full Maxwell equations that will lead to some perhaps surprising results.

Simplify We assume again cylindrical symmetry and we are using a spherical conductor with conductance σ which we assume is high in the sense the effective permittivity is that of a good conductor: see Figure 4.12. Outside the conductor we assume again the long wavelength approximation since this is more common in circuit design. We have inside the conductor

$$\nabla^2 \mathbf{A} + \omega^2 \mu \epsilon' \mathbf{A} = -\mu \mathbf{J}_i$$

Solve Outside the conductor we have again

$$\nabla^2 \mathbf{A} = 0 \quad \text{with solution} \quad A_z(r) = C + B \ln \frac{r}{R_0}.$$

The magnetic flux density is

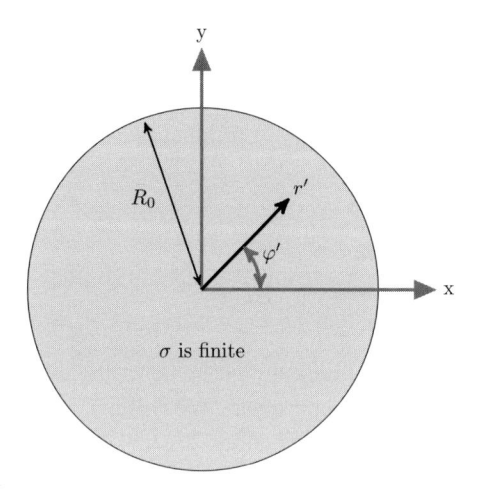

Figure 4.12 Cross-section of single wire with a finite conductance.

$$B_\varphi(r) = -\frac{B}{r}$$

Inside the conductor the general solution is

$$A = \left(\frac{D}{j\kappa} e^{-j\kappa r} + \frac{E}{j\kappa} e^{j\kappa r}\right) e_z$$

representing an outgoing and an incoming wave. We will assume the outgoing wave is zero ($E = 0$). Above

$$\kappa = \sqrt{\omega^2 \mu \epsilon'}$$

With

$$\epsilon' = -\frac{j\sigma}{\omega}$$

We have

$$\sqrt{-j} = \frac{1-j}{\sqrt{2}}$$

We find

$$\kappa = \sqrt{\omega^2 \mu \epsilon'} = \sqrt{-\omega^2 \mu \frac{j\sigma}{\omega}} = \sqrt{\omega\sigma\mu}\,\frac{1-j}{2}$$

We find for the solution

$$A = \frac{D}{j\kappa} e^{-j\kappa r} e_z = \frac{D}{j\kappa} e^{-j\sqrt{\omega\sigma\mu}\frac{1-j}{\sqrt{2}}r} e_z = \frac{D}{j\kappa} e^{-j\sqrt{\frac{\omega\sigma\mu}{2}}r - \sqrt{\frac{\omega\sigma\mu}{2}}r} e_z.$$

4.5 Various High Frequency Phenomena

We see a term in the exponent of the solution that is real and shows a decline as a function of r with a length scale

$$\delta = \sqrt{\frac{2}{\omega\sigma\mu}}. \tag{4.56}$$

Verify This is known as the skin depth of the conductor.

Evaluate The depth to which the electromagnetic fields can penetrate a conductor, skin depth, is dependent on frequency, conductivity and permeability according to (4.35). In typical small geometry CMOS processes the metal layers are really thin – 90 nm is not unusual. For a 6 GHz frequency in copper with a conductivity of $1.67 \cdot 10^{-8}$ ohmm we find a skin depth of 800 nm, 9× the thickness of some metal layers. So, when creating metal shields beware that there are enough metal layers stacked to make an efficient shield given the particular skin depth. Also, the resistance of the metal layers can be substantial – something to keep in mind for large currents.

Another interesting point here is the fact that when the skin depth is less than the conductor size the magnetic field is excluded from part of the conductor. Outside the conductor the field behaves independently of the current distribution inside the conductor and is not affected, as we noted earlier. This exclusion of the magnetic field thus lowers the total magnetic field over all space and the resulting inductance is *reduced*. This effect can be observed in simulators and while often not very big can reduce the inductance by a small percentage.

Key Concept

The depth to which the electromagnetic fields can penetrate a conductor, skin depth, is dependent on frequency, conductivity and permeability according to equation (4.56).

Currents Induced in Perfectly Conducting Ground Plane

We examined this situation previously in section "First Principle Calculation of Single Wire over Ground Plane." We will here examine the induced current in the ground plane for this situation: see Figure 4.13. This calculation is important because it shows how one can estimate the current distributions in a more realistic situation. A good understanding of the current distribution is a key factor to understand inductance.

Simplify We examined this situation previously where we noted the resulting inductance is less when the ground plane is present. Here we will use the same approximations as before but we will look at the field solution near the ground plane. This situation is very common in the circuit world and it is worthwhile to explore this more.

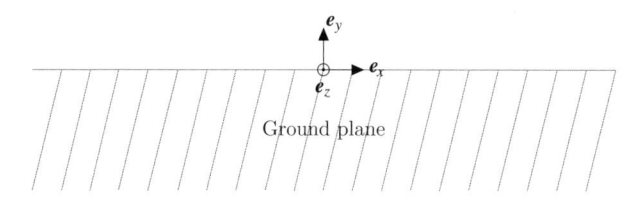

Figure 4.13 Current wire over infinite perfect ground plane.

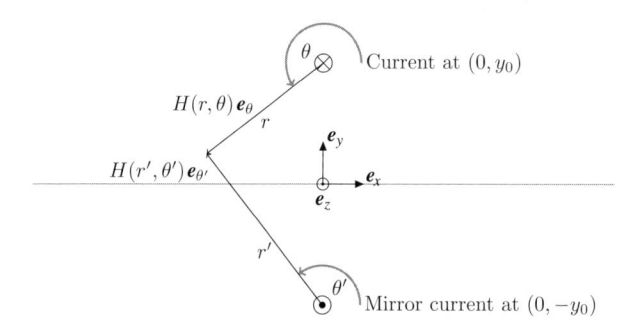

Figure 4.14 Coordinate system of current over ground plane.

Solve The fields above the ground plane can be described as a sum of the field generated from the source current and a mirror field generated from a location equidistant from the ground plane but on the other side of the ground plane. The ground plane itself is removed. The phase of the mirror current is 180 degrees opposite the source current. This is known as the method of images and we described this earlier. We have from Figure 4.14:

$$\mathbf{H}_{sum}(\mathbf{r},\mathbf{r}') = \frac{J_{total}}{2\pi r}\left(\cos\theta\,\mathbf{e}_x + \sin\theta\,\mathbf{e}_y\right) - \frac{J_{total}}{2\pi r'}\left(\cos\theta'\,\mathbf{e}_x + \sin\theta'\,\mathbf{e}_y\right)$$

where $r = \sqrt{x^2 + (y-y_0)^2}$, $r' = \sqrt{x^2 + (y+y_0)^2}$.

At the ground plane we have from the boundary conditions that $H_{sum,x}\big|_{y=0}$ is equal to the surface current in the ground plane. From symmetry we have $r = r'$ and we find from $\mathbf{H}$ above

$$\mathbf{J} = \frac{J_{total}}{2\pi\sqrt{x^2+y_0^2}}2\cos\theta\,\mathbf{e}_z = \frac{J_{total}}{2\pi\sqrt{x^2+y_0^2}}2\frac{y_0}{r}\mathbf{e}_z = \frac{J_{total}}{x^2+y_0^2}\frac{2y_0}{2\pi}\mathbf{e}_z.$$

We see the induced ground current varies with a scale corresponding to the distance of the current above the ground plane: see Figure 4.15. This is as expected. Note that in this long wavelength approximation no other length scales exist with the exception of skin depth which here is zero. We explore this further in the next section.

4.5 Various High Frequency Phenomena

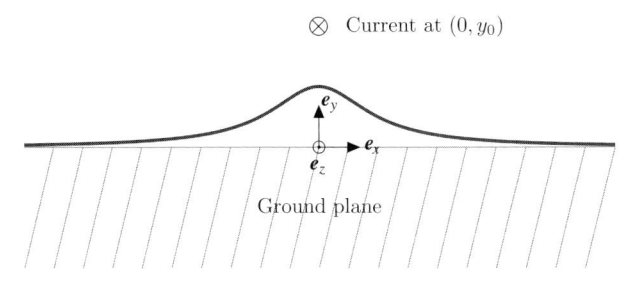

Figure 4.15 Current distribution plot.

Let us integrate this current density in the x-direction

$$\int_{-\infty}^{\infty} J\, dx = \int_{-\infty}^{\infty} \frac{J_{total}}{x^2 + y_0^2} \frac{2y_0}{2\pi}\, dx = 2y_0 J_{total} \int_{-\infty}^{\infty} \frac{1}{x^2 + y_0^2}\, dx = \frac{2y_0 J_{total}}{2\pi y_0^2} \int_{-\infty}^{\infty} \frac{1}{\frac{x^2}{y_0^2} + 1}\, dx$$

$$= \frac{2y_0 J_{total}}{2\pi y_0} \int_{-\infty}^{\infty} \frac{1}{\frac{x^2}{y_0^2} + 1} \frac{dx}{y_0} = \frac{2J_{total}}{2\pi}\left[\tan^{-1}\frac{x}{y_0}\right]_{-\infty}^{\infty} = \frac{2J_{total}\pi}{2\pi} = J_{total}$$

so the total induced current is the same as the forced current. This makes sense since the magnetic field far away will then be canceled out by the two opposing currents.

Verify This is a well-known calculation and can be found in for example [17].

Evaluate The integrated induced current density is equal to the total imprinted current when no current loops are present. If this was not the case there would net magnetic fields at far distances and thus infinite impedance.

> **Key Concept**
>
> The integrated induced current density is equal to the total imprinted current when no other current loops are present.

Currents Induced in Resistive Ground Plane

When solving problems following the estimation analysis method it is not necessary to get an answer valid for all situations. Often it is very instructive to look at extreme points and from those behaviors draw conclusions about the full solution. Our next example will illustrate this point.

We have showed various approximations and how inductance can be derived and estimated depending on the geometry using various approximations that are helpful in integrated circuit design. We owe it to the reader to also present the full solution valid for any wavelength. This is also motivated by the rapid increase in extreme high-speed circuit development. Our discussion is similar to [1].

Simplify Using the full Maxwell equations above we now assume with the same geometry as in Figure 4.13 but the perfect ground plane is replaced a resistive plane:

$$\nabla^2 \mathbf{A} + \omega^2 \mu \epsilon' \mathbf{A} = -\mu \mathbf{J}_i$$

where $\mathbf{J}_i$ is the impressed current. With the chosen coordinate system we find

$$\mathbf{J}_i = J_0 \delta(x)\delta(y - y_0) \mathbf{e}_z, \quad \mathbf{A} = A \mathbf{e}_z.$$

We have

$$\nabla^2 \mathbf{A} + \omega^2 \mu \epsilon' \mathbf{A} = -\mu J_0 \, \delta(x)\delta(y - y_0).$$

Solve Since x is unbounded we can Fourier transform:

$$\delta(x) = \frac{1}{2\pi} \int e^{ix\beta} d\beta \quad A(x, y, z) = \frac{1}{2\pi} \int \tilde{A}_z(\beta, y, z) e^{ix\beta} d\beta.$$

We have

$$\int \left(\frac{\partial^2}{\partial y^2} \tilde{A}_z - \beta^2 \tilde{A}_z + \omega^2 \epsilon' \mu \, \tilde{A}_z \right) e^{ix\beta} d\beta = -\int \mu J_0 \, \delta(y - y_0) e^{ix\beta} d\beta.$$

We can just look at the integrand:

$$\frac{\partial^2}{\partial y^2} \tilde{A}_z - \beta^2 \tilde{A}_z + \omega^2 \epsilon' \mu \, \tilde{A}_z = -\mu J_0 \delta(y - y_0).$$

In general this equation cannot be solved exactly but needs a numerical solution.

For our case the solution has the general form:

$$\tilde{A}_z(\beta, y) = C \frac{j}{2\kappa} e^{j\kappa |y - y_0|} + B e^{j\kappa y}$$

With

$$C = -\mu J_0$$

Inside the lower material the solution is:

$$\tilde{A}_z(\beta, y) = A e^{-j\kappa_2 y}$$

where

$$(-\beta^2 + \omega^2 \epsilon' \mu) = \kappa^2$$
$$(-\beta^2 + \omega^2 \epsilon'_2 \mu) = (\kappa_2)^2.$$

4.5 Various High Frequency Phenomena

The boundary conditions are now

$$\begin{cases} E_z|_1 = E_z|_2 \\ B_z|_1 = B_z|_2 \end{cases}$$

Also

$$B_y|_1 = B_y|_2$$

We get

$$B_x = \frac{\partial \tilde{A}_z}{\partial y}$$

$$B_y = \frac{\partial \tilde{A}_z}{\partial x} = j\beta \tilde{A}_z$$

$$B_z = 0$$

From E_z, B_z is continuous we get

$$\begin{cases} C\dfrac{j}{2\kappa} e^{j\kappa |y_0|} + B = A \\ -\kappa C\dfrac{j}{2\kappa} e^{j\kappa |y_0|} + \kappa B = -\kappa_2 A \end{cases}$$

We see continuity of B_y is guaranteed by the $E_z = E_z$ condition. Multiplying first by κ_2 and adding.

$$C\frac{j}{2\kappa} e^{j\kappa |y_0|} (\kappa_2 - \kappa) + B(\kappa_2 + \kappa) = 0$$

$$B = \frac{\kappa - \kappa_2}{\kappa_2 + \kappa} C \frac{j}{2\kappa} e^{j\kappa |y_0|}$$

and

$$A = C\frac{j}{2\kappa} e^{j\kappa |y_0|} \left(1 + \frac{\kappa - \kappa_2}{\kappa_2 + \kappa}\right) = C\frac{j}{2\kappa} e^{j\kappa |y_0|} \frac{2\kappa}{\kappa_2 + \kappa}$$

We have

$$\tilde{A}_z(\beta, y) = C\frac{j}{2\kappa} e^{j\kappa |y-y_0|} + \frac{\kappa - \kappa_2}{\kappa_2 + \kappa} C\frac{j}{2\kappa} e^{j\kappa |y_0|} e^{j\kappa y} = C\frac{j}{2\kappa} \left(e^{j\kappa |y-y_0|} + \frac{\kappa - \kappa_2}{\kappa_2 + \kappa} e^{j\kappa |y_0|} e^{j\kappa y}\right)$$

$$A(x, y) = \frac{1}{2\pi} \int C\frac{j}{2\kappa} \left(e^{j\kappa |y-y_0|} + \frac{\kappa - \kappa_2}{\kappa_2 + \kappa} e^{j\kappa |y_0|} e^{j\kappa y}\right) e^{jx\beta} d\beta.$$

The first term is the definition of Hankel function of degree zero:

$$H_0(k\rho) = \frac{1}{2\pi} \int \frac{j}{2\kappa} e^{j\kappa |y-y_0|} e^{jx\beta} d\beta.$$

Inside the lower material the solution is:

$$\tilde{A}_z(\beta, y) = C\frac{j}{2\kappa} e^{j\kappa |y_0|} \frac{2\kappa}{\kappa_2 + \kappa} e^{-j\kappa_2 y}$$

$$A_2(x, y) = \frac{1}{2\pi} \int C\frac{j}{2\kappa} e^{j\kappa |y_0|} \frac{2\kappa}{\kappa_2 + \kappa} e^{-j\kappa_2 y} e^{jx\beta} d\beta \qquad (4.57)$$

This is a general solution and needs to be evaluated numerically for a given configuration.

Verify However we can learn a few things by looking at extreme cases of key parameters. Let us use conductivity as such an example. Let us assume we have situation described in the previous section with static field and a ground plane with high conductance. We then have

$$\left(-\beta^2 + \omega^2 \epsilon' \mu\right) \approx -\beta^2 = \kappa^2$$

or

$$\kappa = j|\beta|$$

$$\left(-\beta^2 - j\omega\sigma\mu\right) = (\kappa_2)^2.$$

The solution in media 2 is then

$$A_2(x,y) = \frac{1}{2\pi} \int C \frac{1}{2j|\beta|} e^{-|\beta| \, |y_0|} \frac{2j|\beta|}{\kappa_2 + j|\beta|} e^{-j\kappa_2 y} e^{jx\beta} d\beta$$

$$= \frac{1}{2\pi} \int C \, e^{-|\beta| \, |y_0|} \frac{1}{\sqrt{-\beta^2 - j\omega\sigma\mu} + j|\beta|} e^{-j\sqrt{-\beta^2 - j\omega\sigma\mu} \, y} e^{jx\beta} d\beta.$$

To get the current in the medium we need to take the curl of this twice. This corresponds to multiplying by $-\kappa_2^2$. In this extreme case we can limit ourselves to $y = 0$. So we find.

$$J(x, y = 0) = \frac{1}{2\pi} \int C \, e^{-|\beta| \, |y_0|} \frac{\left(\beta^2 + j\omega\sigma\mu\right)}{\sqrt{-\beta^2 - j\omega\sigma\mu} + j|\beta|} e^{jx\beta} d\beta.$$

For high σ we see the β^2 will not dominate the sum for a long while. We can assume this happens when the first exponent $|\beta| \, |y_0|$ is very large so as to dampen out the integrand. We then get

$$J(x, y = 0) \approx \frac{1}{2\pi} \int C \, e^{-|\beta| \, |y_0|} \sqrt{j\omega\sigma\mu} \, e^{jx\beta} d\beta$$

This is a well-known Fourier inverse function and we have

$$J(x, y = 0) \sim \frac{2y_0}{x^2 + y_0^2}.$$

We see we come back to our old solution, so this is encouraging. What about the other extreme where σ is very small? We can still keep the long wavelength approximation both in medium 1, 2 and we find:

$$J(x,y) = \frac{1}{2\pi} \int C \, e^{-|\beta| \, |y_0|} \frac{\left(\beta^2 + j\omega\sigma\mu\right)}{\sqrt{-\beta^2 - j\omega\sigma\mu} + j|\beta|} e^{-j\sqrt{-\beta^2 - j\omega\sigma\mu} \, y} e^{jx\beta} d\beta.$$

In the extreme case where $\sigma = 0$ we simply recover our old static solution where $J = 0$ outside the current carrying conductor. What if σ is simply really small so that when we look at the integrand that term containing σ is only relevant for small β? By analogy with a more common physical situation where have frequency and time and not wavelength and distance we would say only the low frequency spectrum is affected and thus only long timescales will show any impact of σ. Similarly, we can here say only long distances will show the impact of a situation where the conductivity is low. In other words, for high resistance substrates the current distribution under the current carrying conductor will be of the order $1/\sigma\omega\mu$, which is large.

Evaluate Although we have not provided a formal proof of the following Key Concept it is really just a reworking of Ohm's law: More currents flow where the impedance is small.

Key Concept

In summary, the current in the plane under the signal will follow the path of lowest impedance. For high conductive substrates the dominant impedance is inductance and the current will be tightly distributed directly underneath the wire. For high resistance situations the current will spread out in the plane.

Current Distribution in Thin Conductors

Now that we have spent some time looking at various situations where inductance matters and we have learned the guiding principle is to have the currents flow in such a way as to lower the total magnetic energy, we will here investigate what will happen in a thin conducting slab where the thickness is smaller than the skin depth but the width of the wire is larger than the skin depth. Far away from the wire not much can be done, the generated magnetic field is not sensitive to the local distribution of the currents. But the near field is different and the currents will distribute themselves to sit near the short ends of the wire. This way the magnetic field in the symmetry plane through the conductor is zero and the total magnetic energy is smaller. We will quantify this fact in this section.

We will approach this problem using conformal transformations: see [17, 18]. It is a very powerful way to solve Laplace's equation in two dimensions. It can be used not just in electrostatics but in many other physical fields.

This problem has been studied before in great detail: see [16]. We will not provide the exact solution since it is outside the scope of this book, but we will follow the estimation analysis approach we have been discussing.

Simplify Let us consider the current distribution inside a metal slab. The general solution for a rectangular cross-section is difficult to solve analytically and has not been done to this author's knowledge. However, for an ellipsoidal shape there is a known exact solution: [16]. Here we will just outline the model and do some simple

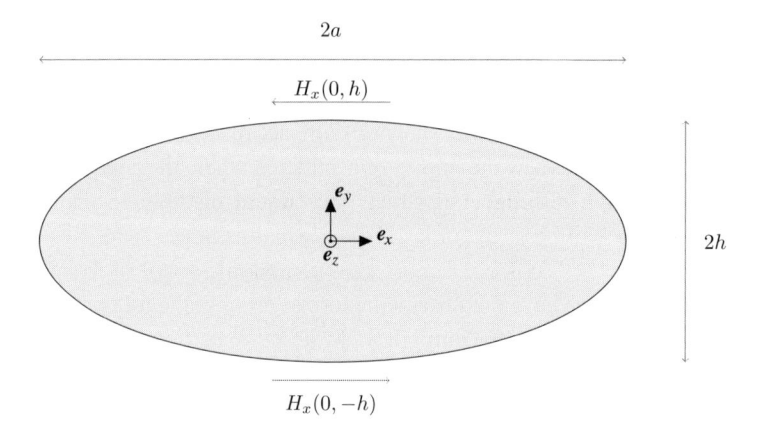

Figure 4.16

Ellipsoidal coordinate system. Adapted from [17].

scaling estimations to arrive at one of the asymptotic solutions. We will use the long wavelength approximation in two dimensions, with the current, J_z, (and vector potential, A_z) going in the z-direction, which states that externally we need to solve Laplace's equation:

$$\Delta A_z = 0 \qquad (4.58)$$

Imagine the conductor has the shape as described in Figure 4.16.

As a function of coordinates x, y the shape is described by

$$y(x) = \pm h\sqrt{1 - (x/a)^2}$$

We will here go to the limit of $h \to 0$ and $\sigma \to$ "large" such that $h \ll \delta \ll a$ and

$$E_z \to 0. \qquad (4.59)$$

For this situation we then naturally have the current in the slab is only dependent on x.

Solve We have from the definitions of the electric field in terms of the gauge potentials (4.49) and our assumption of $E_z = 0$

$$E_z = -\frac{\partial \varphi}{\partial z} - j\omega A_z = 0 \quad \to \quad -j\omega A_z = \frac{\partial \varphi}{\partial z}.$$

Let us take the derivative w.r.t x of the last expression. With our approximations we assume the only current flowing flows in the z-direction. For this to be consistent the voltage drop along the segment in z has to be constant over the surface of the conductor so we expect the derivative w.r.t x to be zero.

$$\frac{\partial(-j\omega A_z)}{\partial x} = -j\omega \frac{\partial A_z}{\partial x} = -j\omega B_y = 0 \qquad (4.60)$$

which becomes our boundary condition at the slab.

4.5 Various High Frequency Phenomena

The problem amounts to solving Laplace's equation for A_z with the boundary condition (4.60) at the conductor. For large distances we must have from (4.49)

$$A_z = -\frac{I\mu}{2\pi} \ln \frac{\sqrt{x^2 + y^2}}{R} \qquad (4.61)$$

corresponding to a current filament with a large return radius R. This is our second boundary condition.

We now introduce the conformal coordinate transformation

$$x + jy = a\cos(u - jv)$$

yielding

$$x = a\cosh v \cos u; \quad y = a\sinh v \sin u \qquad (4.62)$$

$$\frac{\partial x}{\partial v} = a \sinh v \cos u \to 0, \ v \to 0 \qquad \frac{\partial x}{\partial u} = -a \cosh v \sin u \to -a \sin u, \ v \to 0$$

$$\frac{\partial y}{\partial v} = a \cosh v \sin u \to a \sin u, \ v \to 0 \qquad \frac{\partial y}{\partial u} = a \sinh v \cos u \to 0, \ v \to 0$$

The transformation is one-one with the restrictions

$$-\pi \leq u \leq \pi; \quad v \geq 0$$

In the (x, y)-plane the curves of constant v are homofocal ellipses; the segment $y = 0$, $|x| \leq a$ is the infinitely flat ellipse $v = 0$ and v increases outwards to infinity: see Figure 4.17. The curves of constant u are the hyperbolae but there is a cut along the segment $v = 0$, so that u is positive in the upper-half plane Re $y > 0$ and negative in the lower half-plane. The semi-infinite segment $y = 0$, $x \geq a$ corresponds to $u = 0$, whereas the segment $y = 0$, $x \leq -a$ corresponds to $u = \pm\pi$.

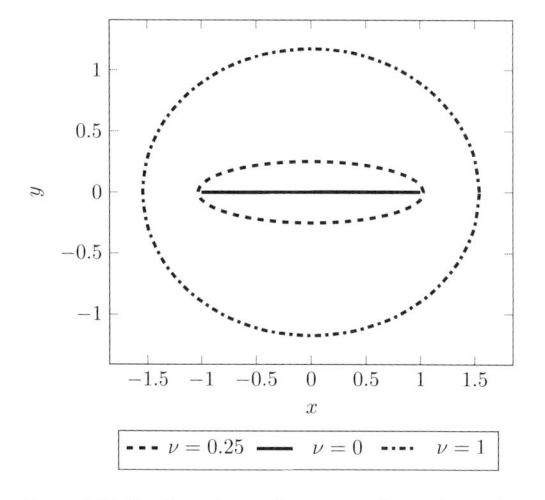

Figure 4.17 Conformal coordinate transformation. Adapted from [17].

With these new coordinates we have for our equation to solve

$$\Delta A_z = 0$$

With the boundary conditions

$$\left.\frac{\partial A_z}{\partial x}\right|_{y\to 0} = \left.\frac{\partial A_z}{\partial u}\frac{\partial u}{\partial x}\right|_{v\to 0} + \left.\frac{\partial A_z}{\partial v}\frac{\partial v}{\partial x}\right|_{v\to 0} = \frac{\partial A_z}{\partial u}\frac{1}{a\sin u} = 0 \to \frac{\partial A_z}{\partial u} = 0$$

The beauty of conformal transformations is if the solution solves Laplace's equation in the new coordinates it also solves it in the old coordinates: see [17, 18]. The main advantage of this transformation is that if the new coordinates follow the basic geometry the equations are easier to solve.

We will further limit ourselves to the situation where $u \neq 0, \mp\pi$ in other words we avoid the endpoints of the ellipsoid. We know the behavior of A_z for large values of v and we also know when $v \to 0$ A_z cannot have any dependence on x (or u in our new coordinates at the limit $v \to 0$). To meet Laplace's equation we know a solution of the form

$$A_z = C + B\,v \qquad (4.63)$$

will work. To find C, B we can attempt a look at the far distance requirement, $v \to \infty$, equation (4.61) and see if we get a self-consistent solution. We have

$$\frac{\sqrt{x^2 + y^2}}{R} \approx \frac{\sqrt{(\cosh v)^2}}{R} = \frac{\cosh v}{R} \approx \frac{e^v}{R}$$

Taking the logarithm of this we find

$$\ln\frac{\sqrt{x^2 + y^2}}{R} \approx v - \ln R$$

We can now identify the constants B, C from (4.61) and (4.63) and we find

$$A_z = -\frac{I\mu}{2\pi}\left[\ln\frac{2R}{a} - v\right] \qquad (4.64)$$

This equation solves Laplace's equation and as $v \to 0$ it has no dependence on u which we need for the boundary condition at the slab and has the right scaling for large v. In other words, it solves the conditions at the boundaries and the equation inside the domain. It is thus a solution to the problem. We will confirm this later when comparing to a more general solution.

We are interested in calculating the current across the thin slab with this solution. It can be calculated from the boundary condition discussed in section "Solutions to Maxwell's Equations: Boundary Conditions," in particular (4.39) where we here find

$$\left.H_x\right|_{y=0+} - \left.H_x\right|_{y=0-} = -H_{x,0} - H_{x,0} = -2H_{x,0} = J_z \qquad (4.65)$$

where $H_{x,0}$ refers to the size of the magnetic field at $y = 0+$. The magnetic field is given by (4.6) and (4.7).

$$H_x\Big|_{y=0+} = \frac{1}{\mu}\frac{\partial A_z}{\partial y}\Big|_{y=0+} = \frac{1}{\mu}\frac{\partial A_z}{\partial y}\Big|_{y=0+} = \frac{1}{\mu}\frac{1}{a|\sin u|}\frac{\partial A_z}{\partial v}\Big|_{v=0+}.$$

We then find the current

$$J_z(u) = -2\frac{1}{\mu}\frac{1}{a|\sin u|}\frac{\partial A_z}{\partial v}\Big|_{v=0+} = -\frac{1}{\mu}\frac{1}{a|\sin u|}\frac{l\mu}{2\pi} = -\frac{I}{a|\sin u|}\frac{1}{2\pi}$$

And in terms of x

$$J_z(x) = -\frac{I}{\sqrt{1-x^2/a^2}}\frac{1}{2\pi}, \quad |x| < a \quad (4.66)$$

Verify This problem has been solved more generally in [17]. The $\sigma \to \infty$ solution in that paper is the same as the one derived here. In the field solver we have analyzed the situation with various thicknesses. See Figure 4.18. As the thickness of the metal slab increase it is getting much better at expelling the field, just as we would expect from the simplified model. For thin wires the prediction fits reasonably well, within 10% except close to the end points (the actual edge is excluded because of the singularity in expression (4.66)). As the thickness increases the effect of the expulsion becomes clear and we see the skin depth start to dominate.

Evaluate The lateral skin effect for wide conductors much thinner than the skin depth has a current length scale variation that is dependent on width only, equation (4.66). As the thickness increases the normal skin effect starts to manifest itself.

> **Key Concept**
>
> The lateral skin effect for wide conductors much thinner than the skin depth has a current length scale variation that is dependent on width only: equation (4.66).

4.6 Summary

In this chapter we have learned

- We have defined Maxwell's equations in a form that is useful for integrated circuit applications.
- We have defined the physical concepts of inductance and capacitance.
- To calculate capacitance the voltages need to be known
- To calculate inductance, the currents need their direction to be known.
- We have looked at various examples where we show how a given source current induces a current in neighboring inductors. Moreover, the current distribution

98   **Electromagnetism: Fundamentals**

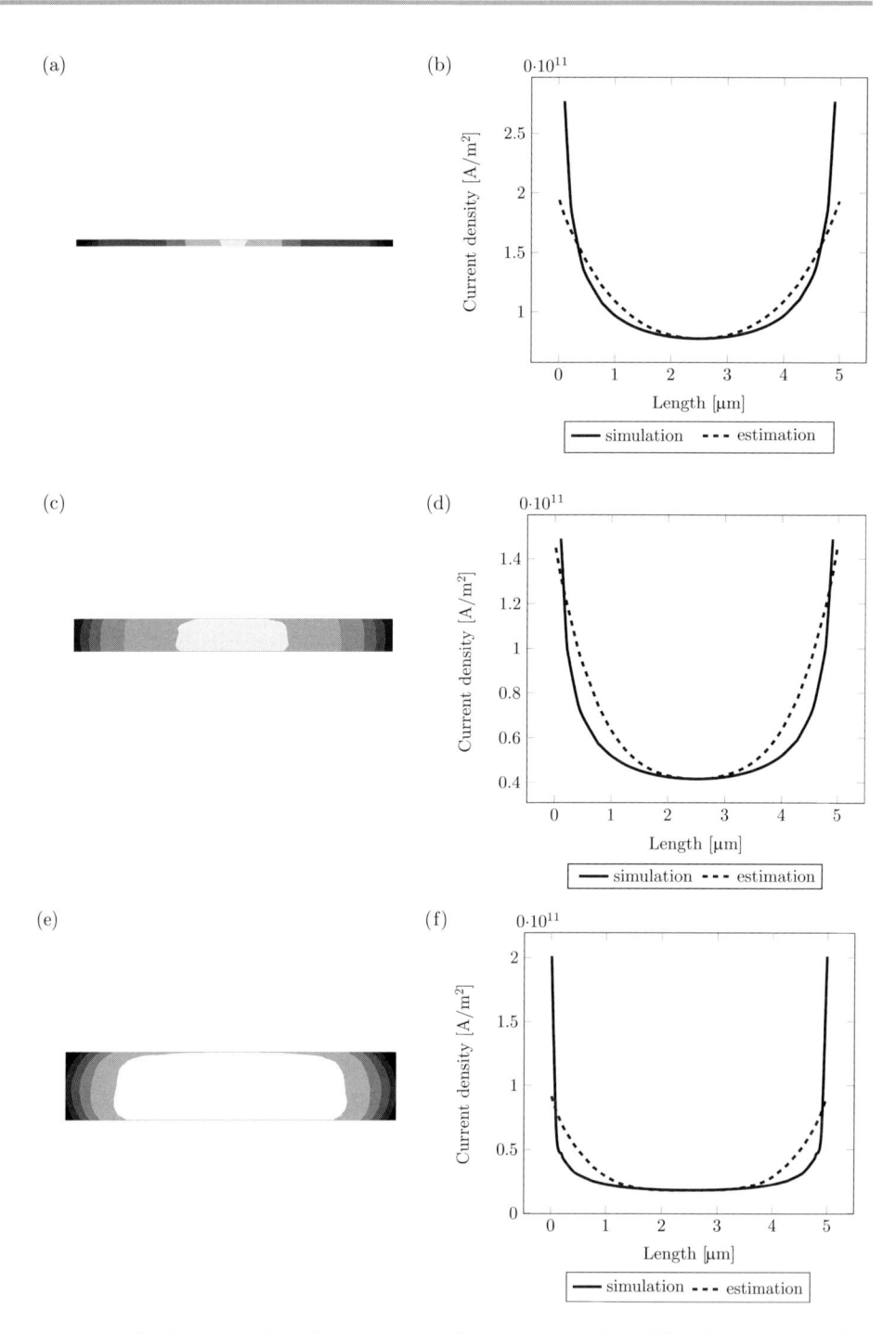

Figure 4.18 Simulation results and comparison wtih estimation analysis. The pictures to the left shows a cross section of the conductor with the current densities indicated in gray. The right pictures shows the current densities through the middle of the conductor. (a, b) corresponds to a thickness of 0.1 μm, (c, d) has a thickness of 0.5 μm, (e, f) shows a slab with 1 μm thickness, (g, h) has 2 μm thickness and finally (i, j) shows 5 μm thickness. The estimated curve has been normalized such that the current density in the middle of the conductor is the same for the simulated and estimated responses.

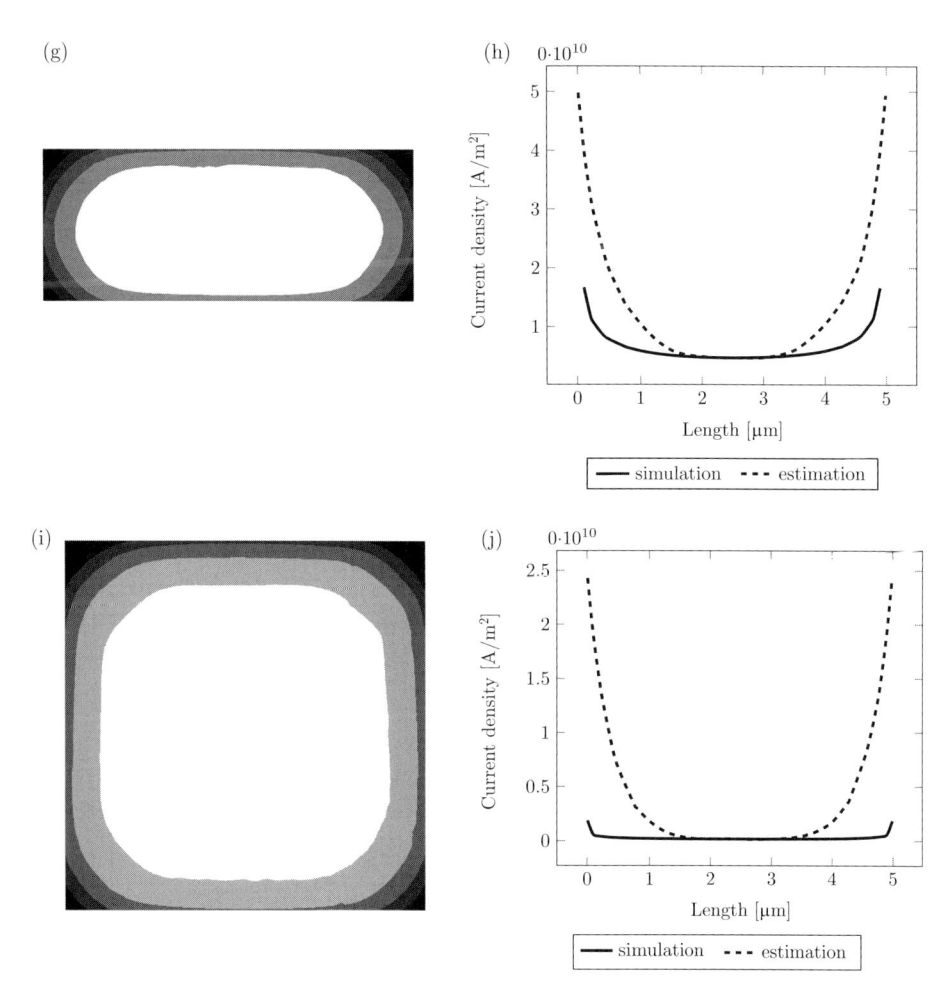

Figure 4.18 (cont.)

within a conductor was derived for a number of examples. When neighboring conductors have significant resistive loss we found a version of ohms law that indicated how the current will distribute itself.

4.7 Exercises

1. Calculate the induced current in an isolated conductor from an outside current. Assume the outside current is fixed. *Hint*: a current sheet with a cylindrical isolated conductor could be a good starting point.
2. The total inductance in section "First Principle Calculation of Two Simple Straight Wires" was calculated assuming the current is evenly distributed inside

the conductors. Calculate the inductance assuming the current only runs on the surface. *Hint*, if one calculates the capacitance the inductance is given by equation (4.55).

4.8 References

[1] H. J. Eom, *Electromagnetic Wave Theory for Boundary Value Problems*, Berlin: Springer Verlag, 2004.
[2] David M. *Pozar, Microwave Engineering*, 4th edn., Hoboken, NJ: Wiley and Sons, 2011.
[3] Malcolm Longair, *Theoretical Concepts in Physics*, Cambridge, UK: Cambridge University Press, 2003.
[4] L. D. Landau, *Lifshitz, Electrodynamics of Continuous Media*, 2nd edn., Oxford, UK: Pergamon Press, 1984.
[5] R. E. Collin, *Foundations for Microwave Engineering*, 2nd edn., IEEE Press on Electromagnetic Wave Theory, Hoboken, NJ: Wiley-IEEE, 1992.
[6] R. F. Harrington, *Time-Harmonic Electromagnetic Fields*, 2nd edn., IEEE Press on Electromagnetic Wave Theory, Hoboken, NJ: Wiley-IEEE, 2001.
[7] Howard W. Johnson and Martin Graham, *High-Speed Digital Design: A Handbook of Black Magic*, Englewood Cliffs, NJ: Prentice Hall, 1993.
[8] L. D. Landau and E. M. Lifshitz, *The Classical Theory of Fields*, Oxford, UK: Pergamon Press, 1984.
[9] A. Shadowitz, *The Electromagnetic Field*, Mineola, NY: Dover Press, 2010.
[10] B. Rojansky and V. J. Rojansky, *Electromagnetic Fields and Waves*, Mineola, NY; J. K. Sykulski, *Engineering Electromagnetism*, Oxford, UK: Oxford University Press, 1994.
[12] J. D. Jackson, *Classical Electrodynamics*, 3rd edn., Hoboken, NJ: Wiley, 1998.
[13] C. A. Balanis, *Advanced Engineering Electromagnetics*, 2nd edn., Hoboken, NJ: Wiley, 2012.
[14] A. Niknejad, *Electromagnetics for High-Speed Analog and Digital Communication Circuits*, Cambridge, UK: Cambridge University Press, 2007.
[15] B. Felsager, *Geometry Particles and Fields*, New York: Springer, 1998.
[16] E. B. Rosa, "The Self and Mutual Inductances of Linear Conductors," *Bulletin of Bureau of Standards*, Vol. 4, No. 2, 1908.
[17] V.Belevitch, "The Lateral Skin Effect in a Flat Conductor," *Philips Technical Review* 32, pp. 221–231, 1971.
[18] H. A. Hermann and J. R. Melcher, *Electromagnetic Fields and Energy*, Englewood Cliffs, NJ: Prentice Hall, 1989.
[19] L. Ahlfors, *Complex Analysis*, New York: McGraw-Hill, 1966.
[20] P. Olver, "Complex Analysis and Conformal Mapping," www-users.math.umn.edu/~**olver**/ln_/cml.pdf, Minnesota, 2017.

5 Electromagnetism: Circuit Applications

> **Learning Objectives**
>
> - Applying estimation analysis to a number of common high-speed situations
>   - Transmission lines
>   - S-parameters
>   - Inductors
>   - Capacitors

5.1 Introduction

This chapter discusses applications of electromagnetism mainly in the context of integrated circuits with the help of estimation analysis. The main purpose of this chapter, as with the other chapters, is to showcase estimation analysis. We will think of ways to simplify a problem, verify that we capture the relevant properties, and then evaluate the result. At the end of the chapter we will use these simplified models in several design examples to establish a starting point for fine-tuning in simulators.

In recent decades, high frequency electromagnetic effects have become increasingly important for on-chip circuit design with the increase in speed and required processing power. Compared with general microwave situations, the length scales under consideration are smaller than typical wavelengths and this provides some natural simplifications for the estimation analyses. Since we are also approaching speeds where the wavelength is comparable to the circuit size scales, we will discuss short wavelength effects in this chapter. Much of what we will be discussing has been touched upon, one way or another, in references [1–18].

We start the chapter with a discussion of the connection between printed circuit board designs and on-die designs. Thereafter we discuss transmission lines and distributed effects. The concept of S-parameters can often be confusing, and we take some time to discuss it from an estimation analysis perspective. This is followed by a section on capacitors, and since this is generally well known, we will be somewhat brief in this discussion. Traditionally they were designed as overlapping metal plates, but with the advance of small geometry CMOS this topology is no longer used much. Instead, MOM (metal–oxide–metal) capacitors where thin, inter-digitized fingers provide the desired capacitance are used. The advancement in lithography has made it possible in modern CMOS technologies to produce such capacitors with much smaller process variation than earlier varieties, and we will briefly discuss the circuit implications. Of course, gate-over-oxide constructions are another approach, and we address this in the section on metal plates. Thereafter we discuss on-chip inductors, where we model various idealized situations and, by removing these idealizations, come to a realistic model of on-chip inductors where capacitance and resistance are also major contributors.

5.2 Connection to PCB Designs

Printed circuit board (PCB) designers have encountered high speed, or distributed, effects for many years not only through fast clock frequencies but also through fast edge rates, and it is only recently that on-die circuit designers have been forced to take such effects into account. On die, the convenient long wavelength approximation was for a long time sufficient for the interconnect modeling. Normally, distributed effects became important in the package interface. Nowadays, these effects have creeped into the on-die interconnects due to the ever increasing need for higher and higher speeds, and a thorough understanding of these kinds of effects has become mandatory. A good description of the problems can be found in [7]. Here many of the electromagnetic properties we found in Chapter 4 are discussed more fully in connection to the PCB design. We will mention just a few of the similarities here and point to what is different, and where different approaches may sometimes be necessary.

Interconnect Scale vs Wavelength

Distributed effects start to become prominent when the wavelength in question, $\lambda = c/f$ where c is the (local) speed of light and f the signal frequency, is similar to the physical length. The important comparison scale turns out to be even smaller than λ and we will derive such results in this chapter. Imagine we have a 10 GHz signal traveling in a uniform medium with a permittivity of 4, we find $\lambda = 1.5 \cdot 10^8 / 10^{10} = 1.5 \, \text{cm}$. On a printed circuit board this is a very small distance. On die, this is fairly large. Ten GHz used to be a respectable frequency, but now imagine a 50 GHz signal. The wave length is 3 mm. A modern high speed integrated circuit can easily be a centimeter or two per side, of which the analog portion is significant. It is then clear that issues such as proper termination to avoid reflections is important.

Ground Planes

These distributed effects were understood very early on in the development of the circuit board, and the concept of proper ground was developed. A modern circuit board can have many tens of layers of copper, each of which is perhaps 15 μm thick, and every second layer is typically used either as a ground or supply plane. On die there can also be more than ten layers, but in contrast to the circuit board system the thickness is only significant (greater than the skin depth) for a few of the top layers. Using lower metals for the ground/return path is woefully inadequate. Instead, co-planar wave guides or multilayer ground planes are commonly used.

Vias

A circuit board via consists of a top/bottom pad and a thin metal cylinder. The pad looks capacitive and the thin cylinder looks inductive, and it can also exhibit distributed effects. An on-die via is simply a stud, often made of tungsten, and its resistance can be significant. One needs to make sure there are plenty of vias when significant current is in play.

Summary

As a short summary of similarities and differences between PCB work and on-chip design, we have:

Similarities:

- Length scale vs wavelength
- Need for proper termination

Differences:

- Thickness of metal layers
- Resistance of vias

5.3 Recent Progress in the Literature on Signal Integrity On-Chip

Over the last several decades there has been a continual need for increased communication speed. Looking at the state of the art a few decades ago, it was not obvious that on-chip inductors would ever be prominent. But the work of the design group under Prof. Robert Meyer at University of California at Berkeley made clear that on-chip inductors have significant advantages. With the ever smaller footprint, the size of inductors can sometimes be prohibitive, but the increase in required speed and thus smaller needed inductances make inductors a key component in high speed designs, where low phase noise oscillators are obvious circuit applications. At speeds in the 100s of GHz range, the problematic passive device is no longer the inductor, which is now very small, but rather the parasitic resistance of capacitors that causes degradation.

This "need for speed" has generated faster and faster integrated circuits, and we are now at the point where the circuit size is of the same order as the wavelength of the signal frequency we are trying to process. The interest in careful modeling of interconnects has likewise increased and several approaches have emerged. For one, accurate and fast simulators that can handle this size and frequency are constantly being developed and improved. Many of the modern simulators have user-friendly interfaces that facilitate setup and simulation. A word of warning to the user, however: oftentimes there are assumptions more or less hidden inside these simulators that may or may not be helpful in solving the problem one is addressing, and one would be wise to spend a good deal of time understanding the inner workings of the tool before embarking on large simulations. Second, the need for accurate analytical understanding has increased in lockstep with simulator developments. Researchers have developed complex models, and we will present an overview of this development in the next section where the focus is on inductor modeling; see also [19] for a recent overview of the field.

Inductor/Interconnect Modeling

Inductors have a high impact on high frequency performance and their modeling is critical to the success of such circuits. One difficulty with inductor modeling is it can take a bit of time to find the right size combination through field solver iterations. In fact, this is one reason that a good initial size estimate, which we will describe later in this chapter, is important. In the literature, the search for accurate inductor modeling is proceeding along several lines of inquiry. One possibility is to use foundry-supplied inductor libraries (see [20–22]), which tends to limit the ability to use an optimal size, but if one finds a reasonable candidate one has access to measured data, which is reassuring from a performance perspective. Another trend is to create various detailed analytical models as in [23–26], π-models as in [27–30], or 2-π models as in [31] to predict inductor performance. There is also [32], where regularization theory is applied to obtain more detailed analytical mapping functions. Another direction was proposed by [33], where machine learning techniques are used to build inductor models. The accuracy is improved by only optimizing promising inductor candidates in EM-solvers instead of simulating each possible inductor, as in [34]. Another approach uses a set of EM-simulated inductors as the design space and the optimal inductor is chosen based on various constraints, as in [35, 36]. The arrival of 3D integrated circuit topologies has generated much work relating to through-silicon-vias (TSV), as in [37–40]. These topologies are there to process data faster in that with the help of TSVs, more data can be transferred vertically, which means a much shorter physical distance and less signal loss is possible.

We have highlighted a handful of situations where several groups have attempted to construct detailed models.

In this chapter we will apply the lessons from Chapter 4 to build useful simple models where the physics of, say, an inductor play a central role. It turns out that having the intuition built from such models is very useful in understanding real situations and can generate a good initial starting point for field solvers.

5.4 Transmission Line Theory

Basic Theory

The focus in this chapter is integrated circuit applications, and here, in general, transmission line effects are not important due to the small length scales involved. However, there are instances, for example inductors, where basic knowledge of transmission line theory is important. We will therefore discuss the basic theory here using estimation analysis, and we refer the reader to the many excellent discussions of the full theory in [2, 5–7, 13].

Simplify A transmission line has essentially two components, one signal conductor and at least one return path as in Figure 5.1.

As we discussed in Chapter 4, this type of structure carries a certain inductance and resistance per length and a certain capacitance to ground per length. In addition, there is a loss in the dielectric medium that we will ignore here. We will model this as a simple RLC filter, as in Figure 5.2. We ignore any loss in the dielectric medium itself, often modeled as a shunt resistor to ground. We are interested in estimating effects such as gain and impedance on-chip, and for common materials inside integrated circuits this loss is negligible.

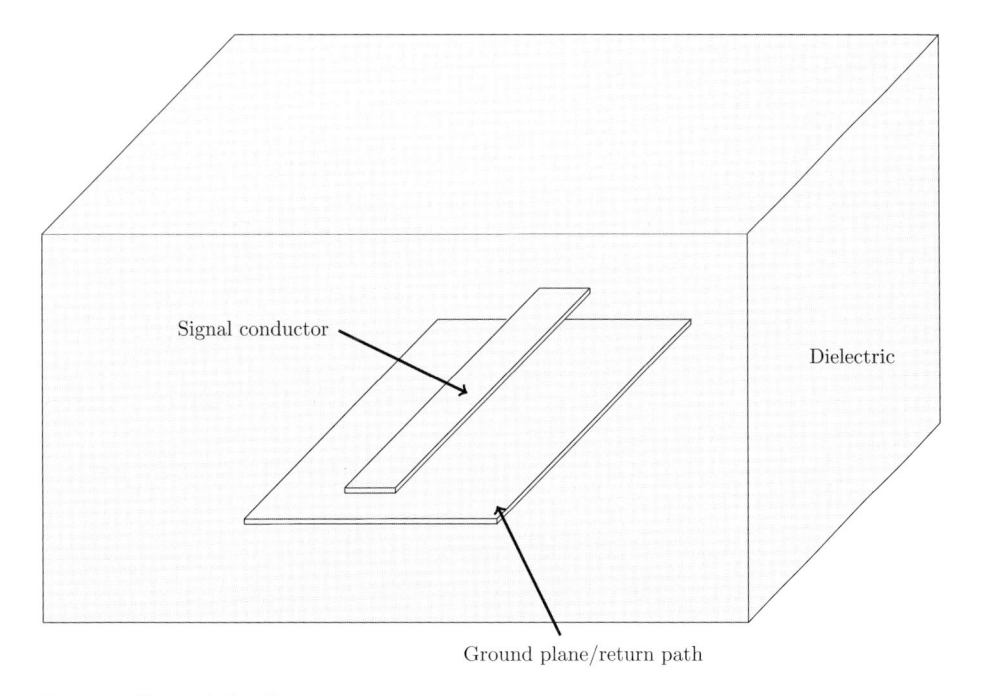

Figure 5.1 Transmission line components.

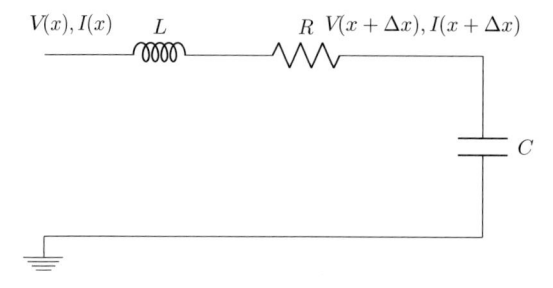

Figure 5.2 Basic transmission line modeling.

Solve We can now analyze the voltages and currents using Kirchoff's current/voltage laws:

$$V(x,t) - R\,\Delta x\, I(x,t) - L\Delta x\, \frac{\partial I(x,t)}{\partial t} - V(x+\Delta x,t) = 0,$$

$$I(x,t) - C\,\Delta x\, \frac{\partial V(x+\Delta x,t)}{\partial t} - I(x+\Delta x,t) = 0.$$

By dividing by Δx and go to the limit $\Delta x \to 0$ we find

$$-\frac{\partial V(x,t)}{\partial x} - R\,I(x,t) - L\frac{\partial I(x,t)}{\partial t} = 0,$$

$$-\frac{\partial I(x,t)}{\partial x} - C\frac{\partial V(x,t)}{\partial t} = 0.$$

We now use the assumption we are in a steady-state condition where the time variation scales as $e^{j\omega t}$. The equations then look like

$$-\frac{dV(x)}{dx} - (R+j\omega L)I(x) = 0, \tag{5.1}$$

$$-\frac{dI(x)}{dx} - j\omega C V(x) = 0, \tag{5.2}$$

Now taking the derivative with respect to x and combining the two we find

$$\frac{d^2 V(x)}{dx^2} = (\omega^2 LC + j\omega RC)\, V(x),$$

$$\frac{d^2 I(x)}{dx^2} = (\omega^2 LC + j\omega RC)\, I(x).$$

The constant in front of the current and voltage terms is known as the propagation constant, $\gamma = \sqrt{(R+j\omega L)j\omega C}$. The solution is well known

$$V(x) = V_0^+ e^{-\gamma x} + V_0^- e^{\gamma x},$$

$$I(x) = I_0^+ e^{-\gamma x} + I_0^- e^{\gamma x}.$$

Where the + refers to a wave going in the positive x-direction and the – refers to a wave going in the negative x-direction. We can use (5.1) above to write the current as

$$I(x) = \frac{\gamma}{R+j\omega L}\left(V_0^+ e^{-\gamma x} - V_0^- e^{\gamma x}\right).$$

We now see we can identify a characteristic impedance as

$$Z_0 = \frac{R+j\omega L}{\gamma} = \frac{R+j\omega L}{\sqrt{(R+j\omega L)j\omega C}} = \sqrt{\frac{(R+j\omega L)}{j\omega C}}. \tag{5.3}$$

We find

$$I(x) = \frac{V_0^+}{Z_0} e^{-\gamma x} - \frac{V_0^-}{Z_0} e^{\gamma x}.$$

Let us now terminate the transmission line with a load Z_L at $x = 0$. We then can define an impedance as a function of x as

$$Z(-x) = \frac{V(-x)}{I(-x)} = \frac{V_0^+ e^{\gamma x} + V_0^- e^{-\gamma x}}{V_0^+/Z_0 e^{\gamma x} - V_0^-/Z_0 e^{-\gamma x}} = Z_0 \frac{V_0^+ e^{\gamma x} + V_0^- e^{-\gamma x}}{V_0^+ e^{\gamma x} - V_0^- e^{-\gamma x}}.$$

At $x = 0$ we get

$$Z(0) = Z_L = Z_0 \frac{V_0^+ + V_0^-}{V_0^+ - V_0^-}.$$

We now find

$$V_0^- = V_0^+ \frac{Z_L - Z_0}{Z_L + Z_0} = V_0^+ \Gamma.$$

Γ is known as the reflection coefficient. We can now write

$$Z(-x) = Z_0 \frac{V_0^+ e^{\gamma x} + V_0^- e^{-\gamma x}}{V_0^+ e^{\gamma x} - V_0^- e^{-\gamma x}} = Z_0 \frac{(Z_L + Z_0)e^{\gamma x} + (Z_L - Z_0)e^{-\gamma x}}{(Z_L + Z_0)e^{\gamma x} - (Z_L - Z_0)e^{-\gamma x}}. \tag{5.4}$$

For $R = 0$ this simplifies to

$$Z(-x) = Z_0 \frac{Z_L \cos \omega \sqrt{LC} x + jZ_0 \sin \omega \sqrt{LC} x}{Z_0 \cos \omega \sqrt{LC} x + jZ_L \sin \omega \sqrt{LC} x}$$

$$= Z_0 \frac{Z_L + jZ_0 \tan \omega \sqrt{LC} x}{Z_0 + jZ_L \tan \omega \sqrt{LC} x}. \tag{5.5}$$

Where we have complied with the norm in the literature and refer to x as $-x$. Let us now look at the special case where $Z_L = 0$ and $R = 0$. We have $Z_0 = \sqrt{L/C}$

$$Z(-x) = j\sqrt{\frac{L}{C}} \tan \omega \sqrt{LC} x, \tag{5.6}$$

which approaches

$$Z(-x) \to j\sqrt{\frac{L}{C}}\omega\sqrt{LC}x = j\omega Lx, \quad \omega x \to 0.$$

We see that when $Z_L \ll Z_0$ the transmission line looks inductive with a total inductance Lx. We need to keep this fact in mind when we discuss inductors, because we must not terminate an inductor at high impedance. We will come back to these equations.

Naturally, for the opposite case where $Z_L = \infty$ we find

$$Z(-x) = \frac{\sqrt{L/C}}{j\tan\omega\sqrt{LC}x} \to \frac{1}{j\omega Cx}, \quad \omega x \to 0.$$

When terminated with an open the transmission line looks like a capacitor.

It is also instructive to see from (5.6) that at

$$\omega\sqrt{LC}x = \frac{\pi}{2} \quad \to \quad x = \frac{c}{4f} = \frac{\lambda}{4}.$$

The impedance goes to infinity. Above this length the impedance changes sign and becomes capacitive in the case of an inductor. This is known as a $\lambda/4$ resonance. There are also resonances at odd integer multiples of this length scale as is clear from (5.6).

CAUTION: One of the difficult parts of microwave engineering is the fact there can be many solutions to Maxwell's equations. For a given boundary there can be many possible modes when the wavelength is similar to the size of the structure. Identifying these modes and removing the unwanted ones is among the major tasks in microwave engineering. For our particular case of a terminated transmission line, sometimes the symmetry or ground plane or other neighboring conductors can cause the electrical length of a $\lambda/4$ resonance to be different from what one would expect from a simple trace length calculation. For instance, a single loop inductor can be seen as a transmission line with its own return path where the halfway point, the termination, is a short. The electrical length will then be half of the physical length of the inductors coil. We will not address the precise root cause of such situations here since it is outside the scope of the book, but we encourage the reader to always verify the resonance location with a simulator. The good news is when the simulator does not agree with a naïve length calculation, it will find a length that is shorter by some integer factor. The resulting resonance frequency is thus higher than one would expect. Here, our calculations will always use the simulated resonance length if it is different from a simple trace calculation.

Verify This is a standard result that, if not expressly given in the many books discussing this subject, can easily be confirmed. See for example [2, 5–7, 13].

Evaluate The impedance of a transmission line depends heavily on its characteristic impedance as well as on how it is terminated. This is most clearly seen in the $\lambda/4$ effect, where the load impedance is effectively inverted when looking from the source point. If one terminates the transmission line with a short it will look inductive when the electrical length is short compared with wavelength, and if one terminates with an open the transmission line will look capacitive.

> **Key Concept**
>
> A transmission line terminated at low impedance, say 0 ohms, looks like an inductor when size ≪ wavelength. A high impedance termination causes the line to look like a capacitor when the size ≪ wavelength.

> **Key Concept**
>
> A $\lambda/4$ resonance is a situation where the length of the conductor is one-quarter of the wavelength. This will transform the impedance at the termination inversely.

Summary

We have applied the estimation analysis to the fundamental concept of a transmission line and demonstrated that we can reach the known solutions with simple mathematical manipulations.

5.5 S-Parameters

Scattering parameters or S-parameters are a fundamental tool in microwave engineering. For the electrical circuit engineer they can often be difficult to conceptualize, and we will show that we can define them in circuit theory (long wavelength approximation) where they are easier to understand. From there a short wavelength extension can be made. We again follow the steps of estimation analysis and build a simple model we can solve to demonstrate some fundamental properties.

We will start with the general definition and refer the interested reader to literature for the details behind them. The discussion follows closely what is often referred to as generalized S-parameters or power waves (see [2]), but here we make the additional assumption that the termination resistor is the same both at the source and the sink, and the degradation is modeled as an impedance between these points. We will then look at a simple circuit version of these parameters and show through some examples how they behave and are related to more familiar concepts such as bandwidth. The last section will discuss the short wavelength generalization.

Definition

From the picture S-parameters are defined as the ratio of incoming or outgoing wave amplitudes in different ports (Figure 5.3).

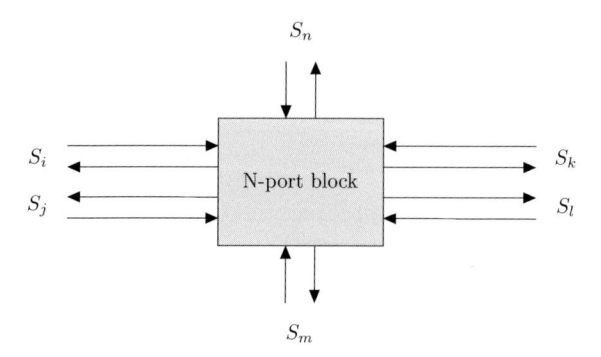

Figure 5.3 Multi-port system showing in-/outgoing waves.

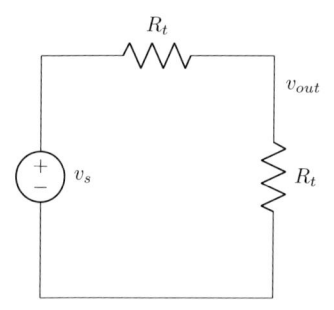

Figure 5.4 A simple model of an input configuration.

For example, S_{11} is the ratio of the outgoing wave amplitude to the incoming wave amplitude when all other ports are 50 ohm terminated. S_{21} is the ratio of outgoing wave amplitude at port 2 divided by the incoming wave amplitude at port 1, all other ports being terminated to 50 ohms. Notice the concept of outgoing and incoming waves: this is a concept with meaning in the short wavelength approximation but it has no natural meaning in the long wavelength limit. Let us keep this in mind in the following discussion.

Simplify We will now simplify the situation to the circuit world or long wavelength approximation. Let us consider the following simple picture (Figure 5.4).

This is a voltage source in series with two resistors. The voltage at point "v_{out}" is clearly $v_s/2$ at all times. What if we change the load resistor by adding a resistor R in series (Figure 5.5)?

Solve Now the output voltage will be

$$v_{out} = \frac{R_t + R}{2R_t + R} v_s.$$

We can manipulate this in the following way

$$v_{out} = \frac{R_t + R}{2R_t + R} v_s = \frac{R_t + R/2}{2R_t + R} v_s + \frac{R/2}{2R_t + R} v_s = 0.5\, v_s + \frac{R/2}{2R_t + R} v_s = S_{in} + S_{out},$$

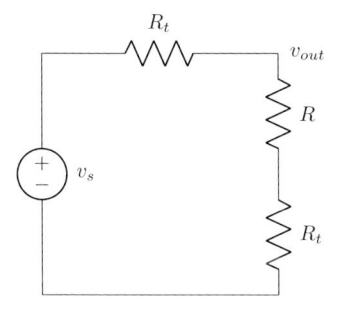

Figure 5.5 A simple model of an input configuration with an additional series resistor.

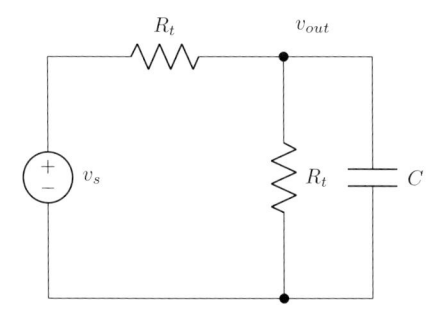

Figure 5.6 A simple model of an input configuration with a shunting capacitor.

where we have defined an "incoming" and "outgoing" wave. In this context S_{11} is simply

$$S_{11} = \frac{S_{out}}{S_{in}} = \frac{|R|/2}{R_t + R/2}.$$

We see that if R is zero, S_{11} is zero. Let us think of a more complicated situation where the load consists of a cap shunting a 50 ohm resistor (as in Figure 5.6):

We find

$$v_{out} = \frac{R_t}{(R_t j\omega C + 1)\left(R_t + \frac{R_t}{R_t j\omega C + 1}\right)} v_s = \frac{1}{((R_t j\omega C + 1) + 1)} v_s$$

$$= 0.5 v_s - \frac{R_t j\omega C/2}{((R_t j\omega C + 1) + 1)} v_s.$$

We see here again we can define S_{11} and find

$$S_{11} = \frac{S_{out}}{S_{in}} = \frac{R_t \omega C/2}{|R_t j\omega C/2 + 1|} = \frac{R_t \omega C/2}{\sqrt{(R_t \omega C/2)^2 + 1}}.$$

For small C we see S_{11} is zero, for large C $S_{11} = 1$. What about S_{11} and bandwidth of the load itself? At bandwidth we have by definition $R_t \omega C = 1$ and we see for this case

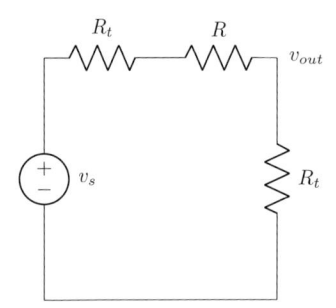

Figure 5.7 A simple model of an input configuration with a series source resistor.

$$S_{11\,BW} = \frac{0.5}{\sqrt{1.25}} = -7\,\text{dB}.$$

How far from the bandwidth do we need to be to get $S_{11} = -20$ dB?

$$\frac{R_t \omega C/2}{\sqrt{(R_t \omega C/2)^2 + 1}} = 0.1 \rightarrow \left(\frac{R_t \omega C}{2}\right)^2 (1 - 0.01) = 0.01 \rightarrow \frac{R_t \omega C}{2} \approx 0.1,$$

which is ~5 times the bandwidth of the unterminated input! The reader should by now appreciate the difficulty in getting low return loss for a given system.

What about the other S-parameters? Let us look at S_{21} using the model in Figure 5.7.

Here we look at the termination as ideal R_t ohms and the driving impedance is different compared with our earlier discussion. We have with a resistive driving impedance

$$v_{out} = \frac{R_t}{R + 2R_t} v_s.$$

The energy that goes into the output load is simply the outgoing wave.

At the input we see the driving impedance as R_t ohms with a load that looks nonideal. With a real nonideality we see

$$v_i = \frac{R_t + R}{R + 2R_t} v_s = \frac{R_t + R/2}{R + 2R_t} v_s + \frac{R/2}{R + 2R_t} v_s.$$

We identify again the first terms as the incoming wave and the last term as the returned amplitude.

We can now identify the various S-parameters:

$$S_{11} = \frac{R}{R + 2R_t},$$

$$S_{21} = \frac{R_t}{R + 2R_t} \frac{1}{1/2} = \frac{2R_t}{R + 2R_t}.$$

(5.7)

We see now

$$S_{11}^2 + S_{21}^2 = \frac{R^2 + R_t^2}{(R + R_t)^2} < 1.$$

5.5 S-Parameters

For this sum to equal one we are missing a term

$$\frac{2RR_t}{(R+R_t)^2}.$$

What happened to it? Since the square of the S-parameters has something to do with power, let us take a step back and think about this situation in terms of power. We have a current

$$i = \frac{v_s}{2R_t + R}.$$

How much power is being burnt in the various resistors?

$$P_{load} = i^2 R_t = \left(\frac{v_s}{2R_t+R}\right)^2 R_t \quad P_{return} = \left(\frac{v_s R/2}{R+2R_t}\right)^2 \frac{1}{R_t} \quad P_R = \left(\frac{v_s}{2R_t+R}\right)^2 R.$$

Let us normalize this to the power burnt in the load termination resistor when $R = 0$.

$$P_{ideal} = \frac{v_s^2}{(2R_t)^2} R_t = \frac{1}{4}\frac{v_s^2}{R_t},$$

We get by normalizing the actual powers in the various resistors to this ideal power,

$$\frac{P_{load}}{P_{ideal}} = \frac{(v_s/(2R_t+R))^2 R_t}{v_s^2/4R_t} = \frac{R_t 4R_t}{(R+2R_t)^2} = \frac{2R_t 2R_t}{(R+2R_t)^2},$$

$$\frac{P_{return}}{P_{ideal}} = \frac{\left(\frac{v_s R/2}{R+2R_t}\right)^2 \frac{1}{R_t}}{v_s^2/4R_t} = \frac{4(R/2)^2}{(R+2R_t)^2} = \frac{R \cdot R}{(R+2R_t)^2},$$

$$\frac{P_R}{P_{ideal}} = \frac{(v_s/(2R_t+R))^2 R}{v_s^2/4R_t} = \frac{R \cdot 2R_t}{(R+2R_t)^2} = \frac{2R \cdot 2R_t}{(R+2R_t)^2}.$$

The last equation shows the missing term from the equation above. It is simply the power burnt in the parasitic resistor R compared with the power burnt in one termination resistor in an ideal situation. The rest of the terms correspond to the power burnt in the load resistor and the power burnt in the source resistor by the return wave.

> **Key Concept**
>
> When the interconnection is resistive it will burn power, so the sum of the square of S_{11} and S_{21} is less than one.

What about the partly inductive interface in Figure 5.8?
Here

$$v_{out} = \frac{R_t}{j\omega L + 2R_t} v_s.$$

This energy that goes into the output load is simply the outgoing wave.

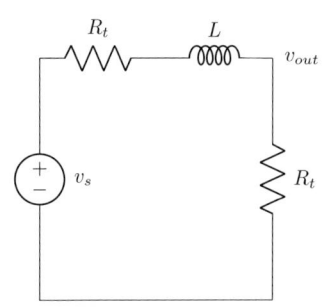

Figure 5.8 An inductor in series with the source impedance.

At the input we see the driving impedance as R_t ohms with a load that looks nonideal. With an inductive nonideality we see

$$v_{out} = \frac{R_t + j\omega L}{j\omega L + 2R_t} v_s = \frac{R_t + j\omega L/2}{j\omega L + 2R_t} v_s + \frac{j\omega L/2}{j\omega L + 2R_t} v_s.$$

We identify again the first terms as the incoming wave and the last term as the returned amplitude.

We can now identify the various S-parameters:

$$S_{11} = \frac{\omega L/2}{\sqrt{(\omega L)^2 + (2R_t)^2}} \frac{1}{1/2},$$

$$S_{21} = \frac{R_t}{\sqrt{(\omega L)^2 + (2R_t)^2}} \frac{1}{1/2} = \frac{2R_t}{\sqrt{(\omega L)^2 + (2R_t)^2}}.$$

We see now

$$S_{11}^2 + S_{21}^2 = \frac{(\omega L)^2 + (2R_t)^2}{(\omega L)^2 + (2R_t)^2} = 1$$

We can also calculate S_{21} for the system with a shunting capacitor. We find

$$S_{21} = \frac{S_{out}}{S_{in}} = \frac{|1/((R_t j\omega C + 1) + 1)| V_s}{0.5 V_s} = \frac{1}{\sqrt{(R_t \omega C/2)^2 + 1}},$$

$$S_{11}^2 + S_{22}^2 = \frac{(R_t \omega C/2)^2}{(R_t \omega C/2)^2 + 1} + \frac{1}{(R_t \omega C/2)^2 + 1} = 1.$$

We can now infer that in a purely reactive environment, the sum of the square of the S-parameters is 1; there is no power lost in the medium. The formal proof is beyond the scope of this book.

5.5 S-Parameters

> **Key Concept**
>
> If the interconnect is purely reactive, the sum of the square of S_{11} and S_{21} is equal to one; there is no power burnt in the interconnection.

For the previous two examples it is obvious that $S_{12} = S_{21}$ due to symmetry. But this is in fact generally true for reciprocal systems (no active devices, ferrites, or plasmas) (see [2]).

Finally, we wrap up this section with an example of how to estimate insertion loss (S_{21}) for a resistive transmission line. Imagine the following situation depicted in Figure 5.9:

We have a transmission line characterized at high frequencies where the skin depth is smaller than the conductor dimensions. It runs over a perfect ground plane (we will generalize this later to a real ground plane).

Simplify We know that the current will run in the bottom of the conductor to minimize the magnetic field, and we know the width of the wire and the skin depth. In Figure 5.10 we have, assuming the dielectric extends beyond the conductor strip,

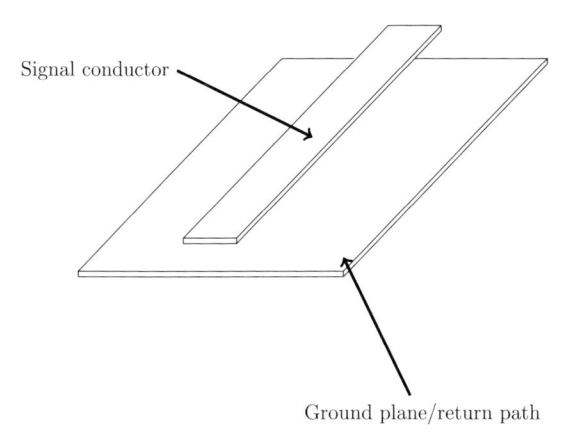

Figure 5.9 Geometry of a single ended transmission line.

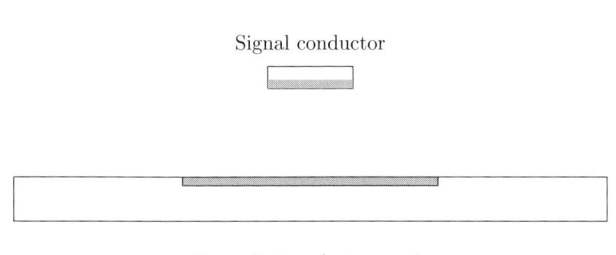

Figure 5.10 A cross-sectional view of the single transmission line.

So
$$R = \rho \frac{L}{sW}.$$

The width of the skin depth in the ground plane is
$$W_{gnd} = W + 2h.$$

Where we have approximated the extension beyond the edge with a generalization of our findings in section "Currents Induced in Perfectly Conducting Ground Plane" in Chapter 4. At first, we will ignore the resistance of the ground plane for simplicity. It will be simple to generalize later.

Solve We now need to calculate the line impedance with the characteristic impedance from (5.3) and a termination impedance $Z_L = R_t$. We get from (5.4)

$$Z = Z_0 \frac{(Z_L + Z_0)e^{\gamma x} + (Z_L - Z_0)e^{-\gamma x}}{(Z_L + Z_0)e^{\gamma x} - (Z_L - Z_0)e^{-\gamma x}}$$

$$= Z_0 \frac{Z_L(e^{\gamma x} + e^{-\gamma x}) + Z_0(e^{\gamma x} - e^{-\gamma x})}{Z_L(e^{\gamma x} - e^{-\gamma x}) + Z_0(e^{\gamma x} + e^{-\gamma x})}$$

We will first look at this from the small electrical length perspective where γx is small.

$$Z = Z_0 \frac{R_t + Z_0 \gamma x}{R_t \gamma x + Z_0}.$$

We now get by using equation (5.3) for

$$Z_0 = \frac{R + j\omega L}{\gamma} \frac{R_t + (R + j\omega L)x}{R_t \gamma^2 x + (R + j\omega L)} \gamma = (R + j\omega L) \frac{R_t + (R + j\omega L)x}{R_t(R + j\omega L)j\omega Cx + (R + j\omega L)}$$

$$= \frac{R_t + (R + j\omega L)x}{R_t j\omega Cx + 1} \approx (R_t + (R + j\omega L)x)(-R_t j\omega Cx + 1)$$

$$= R_t + (R + j\omega L)x - R_t^2 j\omega Cx = R_t + (R + j\omega L)x - j\omega L\, x = R_t + R\,x$$

which is a fairly intuitive result.

We see then from equation (5.7) that

$$S_{21} = \frac{1}{Rx/2R_t + 1}. \qquad (5.8)$$

Verify The transmission line was modeled as a fixed length $\ll \lambda/4$ to avoid distributed effects in line with our previous modeling assumptions. The excitations were modeled as wave ports. From Figure 5.11 we see field solver shows excellent agreement with our theory up to $Rx/R_t \approx 0.1$. After this point there are higher-order effects that start to become important, but we are still within 0.5 dB at $Rx/R_t = 0.5$.

Evaluate The S-parameters can be understood from a simple local picture through the use of generalized S-parameters and power (voltage) waves. Depending on the situation, a number simple scaling laws are easy to derive and confirm.

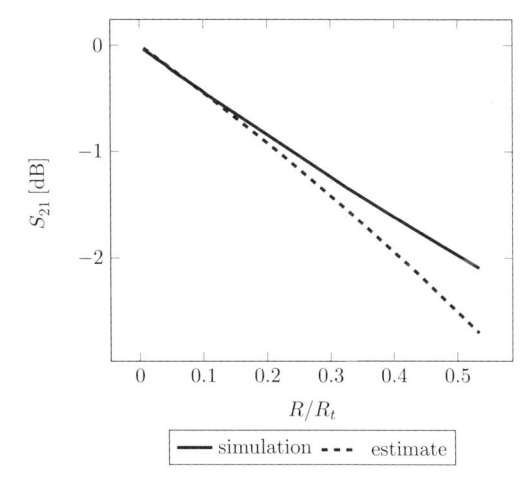

Figure 5.11 A comparison of simulated vs estimation of the loss in a single transmission line.

Summary

There are a couple of convenient microwave theorems useful to keep in mind for passive systems. If the passive system has no resistance, we must have

$$\sum_i S_{i1}^2 = 1.$$

In particular for a two-port system:

$$S_{11}^2 + S_{21}^2 = 1,$$

something we just exemplified.

Another convenient nugget is for a reciprocal system

$$S_{ij} = S_{ji}$$

and for a two-port system

$$S_{21} = S_{12}.$$

S-Parameters for Long Transmission Lines

In the previous section we looked at a short transmission line where the length was much less than a wavelength. Here we will discuss the more general case of arbitrary length t-lines and the resulting S-parameters. This section will tie together a few of the things we discussed earlier in the chapter.

We have the following situation depicted in Figure 5.12.

Simplify We will assume the resistance per unit length, $R \ll \omega L$. We then have for

$$\gamma = \sqrt{(R+j\omega L)j\omega C} \approx j\omega\sqrt{LC}\left(1 + \frac{R}{j2\omega L}\right) = j\omega\sqrt{LC}\,(1+\varepsilon) = j\omega\sqrt{LC} + \frac{R}{2R_t},$$

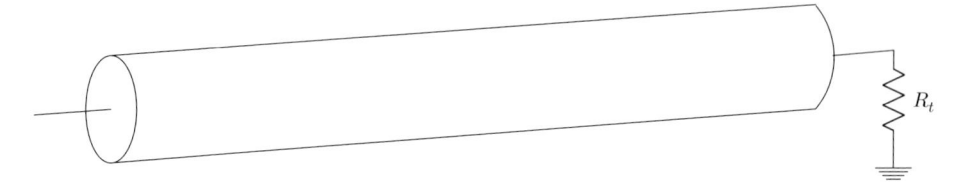

Figure 5.12 Pic of t-line in series with termination resistor.

and

$$Z_L = R_t,$$

$$Z_0 = \frac{R + j\omega L}{\gamma} \approx \frac{j\omega L}{\gamma}\left(1 + \frac{R}{j\omega L}\right) = \frac{j\omega L}{\gamma}(1 + 2\varepsilon)$$

$$= \frac{i\omega L}{i\omega\sqrt{LC}\,(1+\varepsilon)}(1 + 2\varepsilon) \approx R_t(1 + \varepsilon),$$

where

$$\varepsilon = \frac{R}{j2\omega L}.$$

Solve We know from (5.4) that the impedance looking into the transmission line from the source port is

$$Z = Z_0 \frac{(Z_L + Z_0)e^{\gamma x} + (Z_L - Z_0)e^{-\gamma x}}{(Z_L + Z_0)e^{\gamma x} - (Z_L - Z_0)e^{-\gamma x}} = R_t(1+\varepsilon)\frac{(2+\varepsilon)e^{\gamma x} - \varepsilon e^{-\gamma x}}{(2+\varepsilon)e^{\gamma x} + \varepsilon e^{-\gamma x}}$$

$$\approx R_t(1+\varepsilon)\frac{1 + \varepsilon(1 - e^{-2\gamma x})/2}{1 + \varepsilon(1 + e^{-2\gamma x})/2}$$

$$\approx R_t\left(1 + \varepsilon\left(1 - e^{-2\gamma x}\right)\right) = R_t\left(1 + \varepsilon\left(1 - e^{-j2\omega\sqrt{LC}x - Rx/R_t}\right)\right)$$

$$= R_t\left(1 + \varepsilon\left(1 - e^{-Rx/R_t}\left(\cos 2\omega\sqrt{LC}\,x - j\sin 2\omega\sqrt{LC}\,x\right)\right)\right).$$

Following the S_{11} discussion earlier, we have the voltage at the t-line input is

$$V_o = \frac{Z}{R_t + Z} = \frac{Z/2 + R_t/2}{R_t + Z} + \frac{(Z - R_t)/2}{R_t + Z} = S_{in} + S_{out}.$$

where we have defined an "incoming" and an "outgoing" wave. We find

$$S_{11} = \frac{S_{out}}{S_{in}} = \frac{Z - R_t}{Z + R_t}.$$

This expression can be found in most textbooks. Plugging in the numbers we now find

$$S_{11} = \frac{R_t\left(1 + \varepsilon\left(1 - e^{-Rx/R_t}\left(\cos 2\omega\sqrt{LC}\,x - j\sin 2\omega\sqrt{LC}\,x\right)\right)\right) - R_t}{R_t\left(2 + \varepsilon\left(1 - e^{-Rx/R_t}\left(\cos 2\omega\sqrt{LC}\,x - j\sin 2\omega\sqrt{LC}\,x\right)\right)\right)}.$$

Or in terms of magnitude

$$S_{11} = \frac{|\varepsilon|\sqrt{\left(1 - e^{-Rx/R_t}\cos 2\omega\sqrt{LC}\,x\right)^2 + \left(e^{-Rx/R_t}\sin 2\omega\sqrt{LC}\,x\right)^2}}{2}$$

$$= \frac{|\varepsilon|\sqrt{1 - 2e^{-Rx/R_t}\cos 2\omega\sqrt{LC}\,x + e^{-2Rx/R_t}}}{2}.$$

For S_{21} we again look at the voltage at the destination port. We have from Section 5.4

$$V(x) = V_0^+ e^{-\gamma x} + V_0^- e^{\gamma x},$$

where V_0^+ denotes the outgoing wave into the port and V_0^- the reflected wave from the port. We have

$$S_{21} = \frac{V_0^+ e^{-\gamma x}}{V_0^+ e^{-\gamma \cdot 0}} = e^{-\gamma x} = e^{-j\omega\sqrt{LC}\,x - Rx/(2R_t)} = e^{-Rx/(2R_t)}\left(\cos\omega\sqrt{LC}\,x - j\sin\omega\sqrt{LC}\,x\right).$$

In the limit of small x we get

$$S_{21} \approx \left(1 - \frac{Rx}{2R_t}\right)\left(1 - j\omega\sqrt{LC}\,x\right) = \left(1 - \frac{Rx}{2R_t}\right)\left(1 - j\frac{\omega Lx}{R_t}\right) \approx \left(1 - j\frac{\omega Lx}{R_t}\right).$$

We also see that if we keep x small and vary R by changing resistivity of the metal, S_{21} varies with R like

$$S_{21} \sim e^{-Rx/(2R_t)} \approx 1 - \frac{Rx}{2R_t}.$$

Verify This is the same expression we found previously (5.8) in the limit $Rx \ll R_t$.

Evaluate The S-parameters for a long transmission line can be estimated by looking at the impedance at the input of the line and from there calculating the return loss. The through response can be estimated from the results of wave equation directly.

5.6 Capacitors in Integrated Circuits

Capacitors in integrated circuits are oftentimes well understood by engineers. Here we will discuss them very briefly for the sake of completeness. We start with plate capacitors, which were more commonly used in older technologies, and continue with finger capacitors, which are used with modern CMOS processes.

Capacitors in Integrated Circuits: Plate Capacitors

The MIM cap, or metal–insulator–metal capacitor, is an older type of capacitor. It consists of two large plates with a thin insulator between them as shown in Figure 5.13.

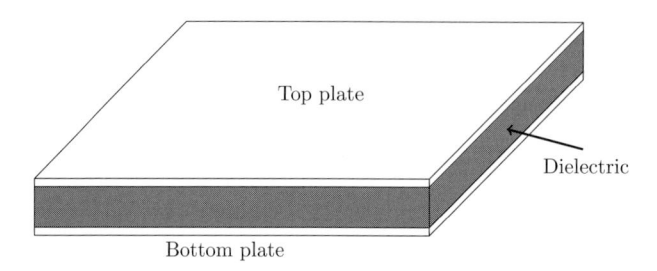

Figure 5.13 Basic capacitor model.

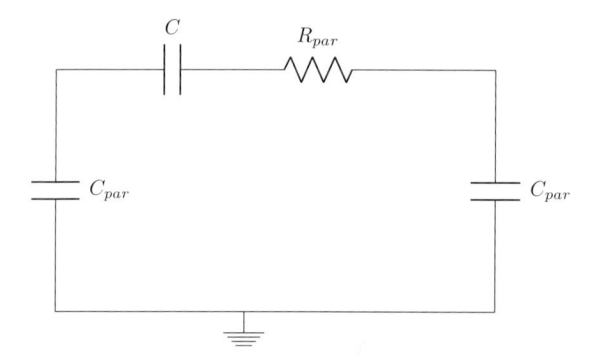

Figure 5.14 Basic circuit model of capacitor.

In modern CMOS processes they are generally no longer available. The closest approximation to a plate capacitor is the MOS cap, where the top plate is the transistor gate and the bottom plate is the transistor body.

Simplify For both of these kinds of capacitors we can approximate the capacitance with the model in Chapter 4, section "Simple Two-Plate System Calculation."

In Figure 5.14, we also ignore the parasitic resistance R_{par}.

Solve We find

$$C_{mim} = \frac{A_{rea}}{d}\varepsilon_0\varepsilon$$

and

$$2C_{par} = \frac{A_{rea}}{d_{sub}}\varepsilon_0\varepsilon$$

where d is the distance between the plates and d_{sub} is the distance to substrate.

Verify This is a well-known approximation that can be found for example in [15, 16].

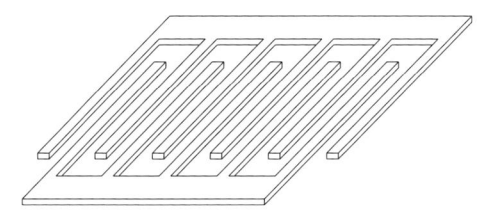

Figure 5.15 Finger capacitor (MOM) configuration, popular in small geometry CMOS. The figure shows one metal layer; often there are several, up to 10!, stacked on top of each other.

Evaluate We have ignored the parasitic resistance in the model for the capacitor. In this case a capacitor can be described as an ideal capacitor with parasitic capacitors to ground.

Capacitors in Integrated Circuits: MOM Capacitors

Modern CMOS technologies have done away with the MIM capacitor layers and instead rely on thin interlocked fingers as in Figure 5.15.

Simplify We can simplify the capacitance calculation by assuming that the sidewall–sidewall capacitance dominates. The parasitic capacitance to ground is simply given by the capacitance area. The circuit model is the same as in Figure 5.14. The parasitic resistance is ignored here as well.

Solve We have

$$C_{mom} = C_{corr} N_{fingers} N_{layers} \frac{lt}{d} \varepsilon_0 \varepsilon \tag{5.9}$$

and

$$2C_{par} = \frac{A_{rea}}{d_{sub}} \varepsilon_0 \varepsilon$$

where $C_{corr} \sim 2$ is correction constant due to topside wall added capacitance, A_{rea} is the total cap area, and d_{sub} is the distance to substrate.

Verify Table 5.1 shows a comparison of estimated and simulated capacitances. The accuracy for these models is less than what is normal for the larger plate capacitors since other metal faces are contributing significantly to the capacitance, hence the C_{corr}.

Evaluate The capacitance of a MOM cap or finger cap scales as the number of fingers times their thickness times their length divided by the distance between the fingers according to (5.9).

Table 5.1 Comparison of estimated vs simulated capacitances for MOM capacitors

C_{mom} with t = 90 nm for various parameters	Estimated ($\varepsilon = 3$) fF	Simulated from a realistic PDK fF
Nlayers = 3, Nfingers = 10, d = 50 nm, l = 1 μm	2.86	2.90
Nlayers = 6, Nfingers = 10, d = 50 nm, l = 1 μm	5.72	5.50
Nlayers = 3, Nfingers = 20, d = 50 nm, l = 1 μm	5.72	6.02
Nlayers = 3, Nfingers = 10, d = 100 nm, l = 1 μm	2.86	2.30
Nlayers = 3, Nfingers = 10, d = 50 nm, l = 2 μm	5.72	5.57

5.7 Inductors in Integrated Circuits

Inductors in integrated circuits have been discussed for many years by many helpful contributors. For example, [14] is now a classic book on inductors and it is also the basis of a very useful software tool, ASITIC, that is widely used. In addition, [15, 17] have good discussions on various aspects of inductors.

We will exemplify the use of the method of estimation by applying it to the inductor circuit element. The discussion starts with a simple wire stub pair, then we use what we have learned and build a complete rectangular inductor using two such pairs. The method of estimation is then applied to find out how one can simply model rectangular cross-sections, which is a somewhat more complicated situation. A more complicated two-turn inductor is encountered after this and the effect of a ground plane on the inductance is thereafter examined. We wrap up the section on inductors by applying estimation analysis to the idea of self-resonance and a more complete model where parasitic resistors and capacitors are included. All these sections will produce useful shorthand formulae that can be used in real world situations, and we show this by applying them to a number of design examples later in the chapter.

Partial Inductance of a Wire Stub Pair

There is of course no such thing as a single wire stub inductor; there needs to be a closed loop somehow. We thus approximate such a loop with a similar wire stub close by, where the current goes the other way as shown in Figure 5.16. In fact, we have already solved a very similar problem in Chapter 4, section "First Principle Calculation of Two Simple Straight Wires." As the reader may recall, we assumed two infinitely long wires and calculated the inductance per length. Here we will use that calculation with a trivial simplification.

Simplify The inductance per length for the infinitely long circular pair was

$$\frac{L}{Z} = \frac{\mu}{\pi}\left(\frac{1}{4} + \ln\left(\frac{d}{R_0}\right)\right)$$

where L/Z is the inductance per length. We will simplify our problem with finite length wires by assuming the magnetic field is independent of position along the wire and

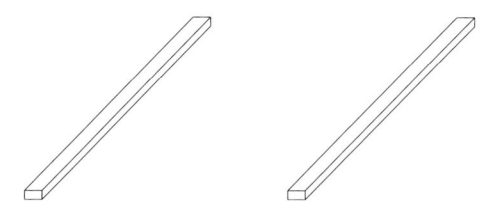

Figure 5.16 "Partial" inductor with two finite length wires.

abruptly ends at the end of the wire. In other words, we assume the field is given by the approximation of Chapter 4, section "First Principle Calculation of Two Simple Straight Wires" along the wire and zero beyond it. We also approximate the rectangular cross-section with a circular one.

Solve The inductance of the wire is now simply

$$L_{pair} = \frac{L}{Z} l$$

Verify In practice, for this situation the field is smoothly reduced towards the end of the wire and is somewhere around 10–15% overestimated with this model.

Evaluate This expression is telling us the inductance is proportional to the length of the wire segment, which is an important lesson. One also sees that the inductance varies with the logarithm of the distance between them, so when estimating inductances the area size of the inductor will also play a role. Imagine a situation where the two segments lie inside each other: clearly the length is there, but since they completely overlap, cancelling each other's current, the total inductance is zero.

One should also note that here we are strictly describing partial inductance since we do not have a full loop. Depending on how the current actually closes in a loop, the total inductance can be very different. We will remedy this situation in the next section.

Inductance of Two Wire Stub Pairs at a Right Angles

Having been through the previous simple example, let us now look at the situation where we have two such pairs that are perpendicular to each other and connecting in such a way that we have a closed rectangular inductor. In this situation, we have four wires carrying current that contribute to the inductance. The pairs need not be equal in size.

We can compare this to a typical on-chip inductor in Figure 5.18.

It is clear from Figure 5.18 that the approximation given by Figure 5.17 is a fairly reasonable one.

Simplify Formally, we now need to follow Chapter 4, section "Field Energy Definitions" and calculate the interaction between the vector potential, A_j, from each current segment, J_j, with all other current segments, J_i. However, it is now clear there is no

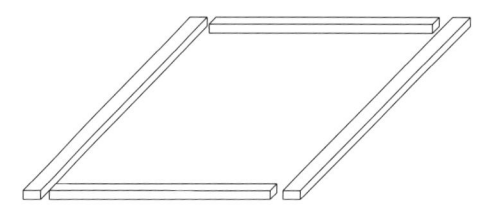

Figure 5.17 Two pairs connecting to each other.

Figure 5.18 A typical on-chip inductor in 2D projection.

interaction between the fields and currents of the two separate pairs. All these integrals are equal to zero since $A_j \cdot J_i = 0$ when i,j belong to different pairs. The simplification we make here is that at the wire ends, the noninteraction between the fields of the two wires is maintained. The total magnetic field energy is thus equal to the sum of the energy from the two individual pairs.

$$|\boldsymbol{B}|^2_{total} = |\boldsymbol{B}|^2_{pair1} + |\boldsymbol{B}|^2_{pair2}.$$

Finally, we simplify the inductance rectangular cross-section with a circular one. The accuracy of this will be addressed in section "An Inductor Model with Rectangular Cross-Section."

Solve The total inductance is now simply the sum of the partial inductances of the two pairs,

$$L = \frac{\mu}{\pi} l_{pair1} \left(\frac{1}{4} + \ln\left(\frac{d_{pair2}}{R_0}\right) \right) + \frac{\mu}{\pi} l_{pair2} \left(\frac{1}{4} + \ln\left(\frac{d_{pair1}}{R_0}\right) \right). \tag{5.10}$$

For the special case where the legs are of equal length, we find

$$L = \frac{2\mu l}{\pi} \left(\frac{1}{4} + \ln\left(\frac{l}{R_0}\right) \right) \tag{5.11}$$

where we have used $d_{pair1} = d_{pair2} = l_{pair1} = l_{pair2} = l$ since the two pairs have equal length. The total length of the structure is $4l$.

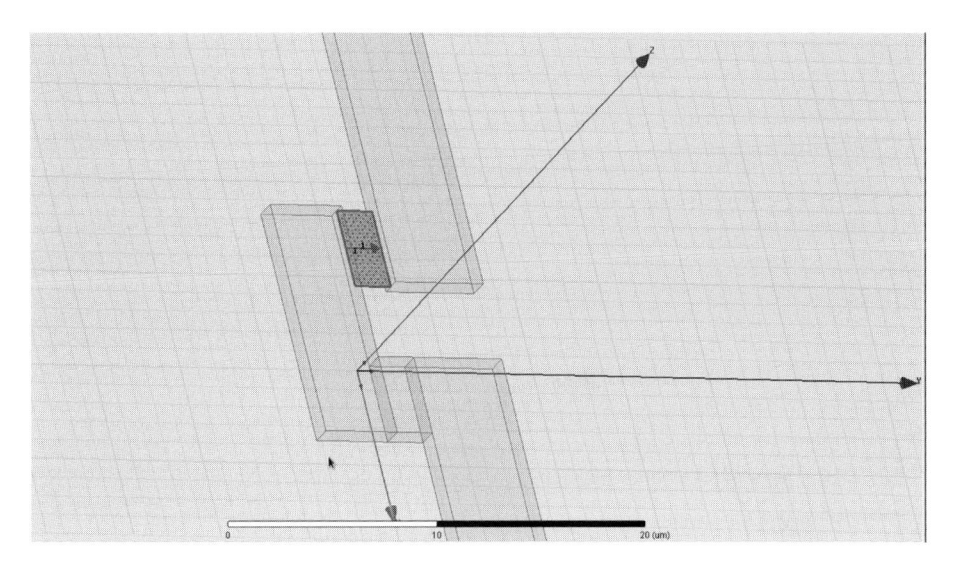

Figure 5.19 Typical field solver setup.

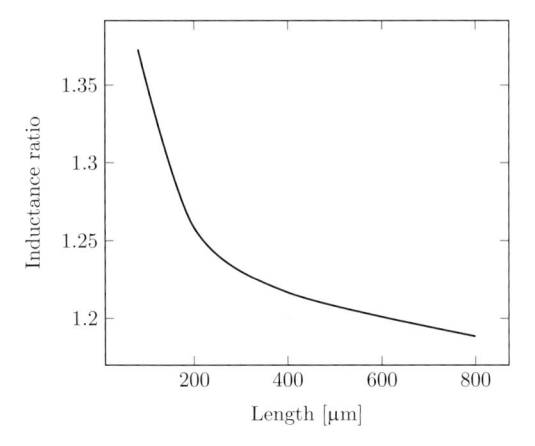

Figure 5.20 On-chip inductance vs length of inductor compared with simulation.

Verify In Figure 5.20 we show a field solver solution of the square inductor structure in Figure 5.18 as a function of l with a fixed width, R_0, and compare it with the approximate solution in equation (5.11).

The field solver setup is as shown in Figure 5.19.

The inductor structure sits in an Air box with radiating boundary conditions. The size of the Air box is about three to four times larger than the inductor itself. The size was shown to be sufficient by making it larger and not seeing any change in the result.

The port excitations are modeled as a lumped port in one end and a short to ground in the other. Ground is modeled as a simple perfect conductor stub closing the loop. The inductance is then found as the imaginary part of the $Z(1,1)$ impedance.

Electromagnetism: Circuit Applications

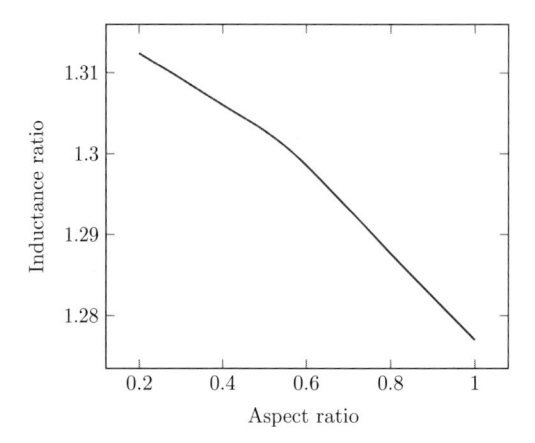

Figure 5.21 On-chip inductance vs aspect ratio of inductor compared with simulation.

$$L_{sim} = \frac{\text{im } Z(1,1)}{\omega}$$

All the inductors discussed in this chapter are modeled along the same lines. The material surrounding the inductor is simply Air with radiating boundary conditions.

We also simulate an inductor with different aspect ratios and find the result in Figure 5.21.

Evaluate The inductance of a single-turn inductor scales roughly linearly with the conductor's length. There is also a logarithmic term that depends on the inductor's length divided by its width. To change the inductance, it is much more important to change its length than its width. In addition, different aspect ratios can be modeled with up to 30% accuracy.

Key Concept

The inductance of a single-turn inductor scales roughly linearly with its length. The width of the inductor is logarithmically dependent on its width.

An Inductor Model with Rectangular Cross-Section

To this point our formulae have approximated the square cross-sections of the conductors with a circular one. As the aspect ratio of the cross-section deviates from a square, the approximation will increasingly deviate from expectation. In this section we will discuss various ways to take rectangular cross-sections into account.

Two Adjacent Circular Segments

One way to extend the previous models to rectangular cross-sections is to put another circular conductor adjacent to the ones we have. In a cross-section we find as shown in Figure 5.22.

5.7 Inductors in Integrated Circuits

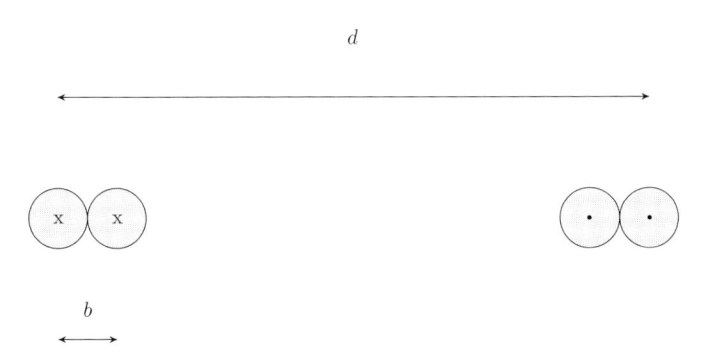

Figure 5.22 Cross-section of an inductor where a 2-1 rectangular cross-section is modeled by two cylinders touching each other.

Simplify The simplification here is simply using two circles as a description of a rectangle.

Solve We can now apply equation (4.52) to this model by noting that instead of one pair of circular conductors there are six (!) pairs, with the current in each conductor equal to half of the total current. This last point is important. The previous discussion assumed the total current in a leg was J. To model the rectangular cross-section, the total current still needs to be J resulting in $J/2$ running in each of the conductors. In the next section we will discuss what happens when we arrange the current such that it is equal to J in each segment. We find here for the energy

$$F_{rect/2pair} = \frac{(J/2)^2}{2}\frac{2\mu}{\pi}d\left(\frac{1}{4} + \ln\frac{d}{H/2} + \frac{1}{4} + \ln\frac{d-2b}{H/2} - 2\ln\frac{b}{H/2} + 2\ln\frac{d-b}{H/2}\right)$$

where H is the thickness of the rectangle. By going to the limit $b = H \ll d$ and again removing $J^2/2$ we find for the inductance

$$L_{rect/2pair} = \frac{1}{4}\frac{2\mu}{\pi}d\left(\frac{1}{2} + 4\ln\frac{d}{R_0} - 2\ln 2\right) = \frac{2\mu}{\pi}d\left(\frac{1}{8} + \ln\frac{d}{R_0} - \frac{1}{2}\ln 2\right)$$

Verify A comparison with simulation of a 200 µm-long, 1 µm-thick, 2 µm-wide inductor shows an $L_{sim} = 134.7$ pH compared with $L_{est} = 175$ pH, an error of 30%.

Evaluate A simple extension to the square model by using two circular segments touching each other such that a 2-1 ratio of conductor width vs length can be modeled shows the same 30% accuracy we had for the square case.

Defining an Effective Radius from Area Equivalence

Another way of modeling a rectangular cross-section ($H \times W$) is to define an effective radius. We will first use an effective radius based on equal area assumption.

Simplify

$$R_0 = \sqrt{\frac{HW}{\pi}}$$

Solve We get from (5.10)

$$L = \frac{2\mu d}{\pi}\left(\frac{1}{4} + \ln\left(\frac{d}{\sqrt{HW/\pi}}\right)\right) \qquad (5.12)$$

For a 2 × 1 rectangle we find

$$L = \frac{2\mu d}{\pi}\left(\frac{1}{4} + \ln\left(\frac{d}{H\sqrt{2/\pi}}\right)\right) = \frac{2\mu d}{\pi}\left(\frac{1}{4} + \ln\left(\frac{2d}{H}\right) - \frac{1}{2}\ln 2 - \frac{1}{2}\ln\left(\frac{4}{\pi}\right)\right)$$

Verify This is very similar to the previous expression. In particular the last term is $\sim 1/8$.

Evaluate The precise model of the "radius" of the conductor matters little as long as the size of the inductor is large compared with this radius. The logarithm is very forgiving when its argument is large to begin with.

Defining an Effective Radius from Max Scale

There are other ways to define an effective radius, and from a practical point of view they can work better.

Simplify Let us define an effective ratio as the max of the sides of the conductor cross-section.

$$R_0 = \text{MAX}(W, H)/2$$

Solve We then have for $W > H$

$$L = \frac{2\mu d}{\pi}\left(\frac{1}{4} + \ln\left(\frac{2d}{W}\right)\right) \qquad (5.13)$$

Verify Using this formula, we see a much improved comparison with simulation in Figure 5.23.

Evaluate The reason is simply that two wrongs make a right. The formula significantly overestimates the effective radius of the conductor cross-section, which produces a smaller logarithmic term. However, the field strength is already overestimated due to the finite size of the length of the conductor. These two errors counter each other and produce results that are closer to the simulations. The formula will obviously work very poorly for extreme cross-section aspect ratios, but keeping it to 1/10 or less as in

5.7 Inductors in Integrated Circuits

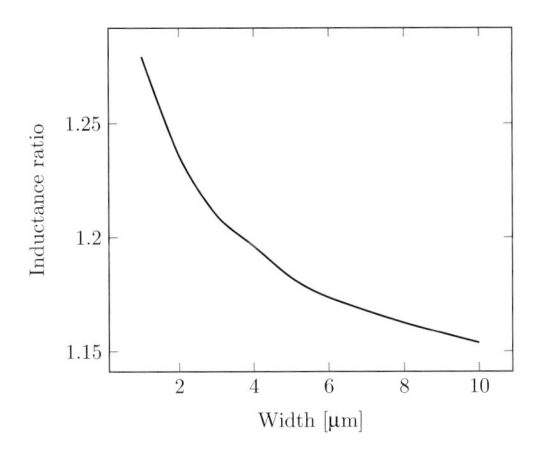

Figure 5.23 Estimation vs simulation of inductance vs width using equation (5.13).

Figure 5.23 shows we are in a comfortable region. This is also well covered by typical integrated circuits applications.

Summary – One-Turn Inductor
In the previous four subsections we applied estimation analysis to a one-turn inductor. We saw that by thinking through what a real inductor looks like and comparing it with problems we already solved in Chapter 4, we could come up with a really simple model that catches many of the aspects of the physical structure and come within 10–30% of the actual value of inductance. This is a good example of efficient application of estimation analysis. We will continue this method of estimation by applying it to a two-turn inductor.

A Model of a Two-Turn Inductor

We can take this model one step further and look at a two-turn inductor as pictured in Figure 5.24.

Simplify We simplify this by following the strategy in the previous section, but here we will allow the second conductor to have some space between it and the outer turn: see Figure 5.25.

We further assume the internal pair has length $l_{int} = d - 2b$ and that the interaction with the other conductors is limited to the same length.

Solve We can now use equations (4.52) to take all the interactions between the conductors into account. By noting that contrary to the rectangular cross-section discussion in section "Two Adjacent Circular Segments" the full current J now goes through each of the legs, we simply feed the current back in a second loop. We find for the energy

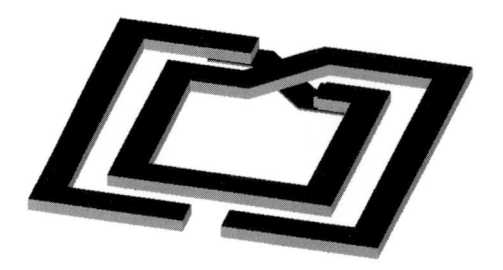

Figure 5.24 Picture of a two-turn inductor.

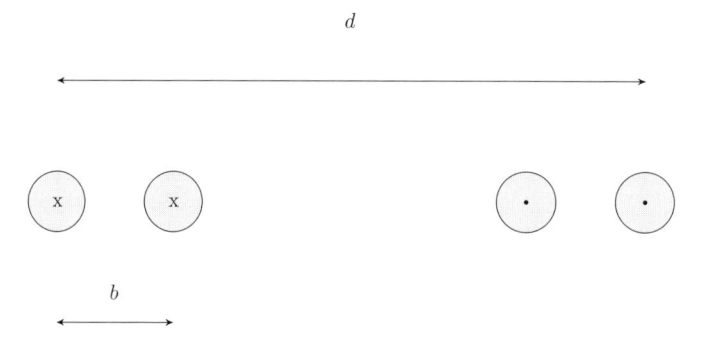

Figure 5.25 Picture of a cross-section of a two-turn inductor.

$$F_{2\text{-}turn} = \frac{2\mu J^2 d}{2\pi}$$
$$\times \left(\frac{1}{4} + \frac{1}{4}\frac{d-2b}{d} - \frac{2(d-2b)}{d}\ln\frac{b}{R_0} + \frac{2(d-2b)}{d}\ln\frac{d-b}{R_0} + \frac{(d-2b)}{d}\ln\frac{d-2b}{R_0} + \ln\frac{d}{R_0} \right)$$

The corresponding inductance, after some rearrangement, is

$$L_{2\text{-}turn} = \frac{2\mu d}{\pi} \left(\frac{1}{4} + \ln\frac{d}{R_0} + \frac{d-2b}{d}\left(\frac{1}{4} - 2\ln\frac{b}{R_0} + 2\ln\frac{d-b}{R_0} + \ln\frac{d-2b}{R_0} \right) \right) \quad (5.14)$$

We now see as $b \to 2R_0$ where $R_0 \ll d$

$$L_{2\text{-}turn}|_{b=2R_0 \ll d} = \frac{2\mu d}{\pi} \left(\frac{1}{2} + 4\ln\frac{d}{R_0} - 2\ln\frac{b}{R_0} \right) \quad (5.15)$$

We compare this to the previous calculation, equation (5.11), and see we have ~4× the inductance when $d \gg R_0$. In effect, looking at the inductance in terms of magnetic energy, we find for this case that the magnetic field is doubled and the energy has quadrupled. In a circuit context this is often discussed in terms of inductive coupling between the coils.

Verify In Figure 5.26 a real two-turn inductor with rectangular cross-section is modeled and compared with equation (5.14) with various combinations of b, d. The field solver setup is similar to what was done for the single-turn inductor, where we

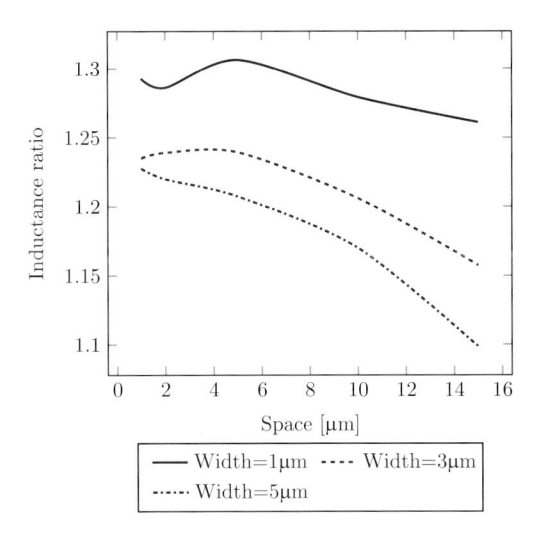

Figure 5.26 Comparison with simulation of the estimated inductance.

evaluate the simulated inductance in the same way. The estimations for various widths use equation (5.13).

Evaluate We see from the result that the error is less than some 30% for the inductance, which implies the average field amplitude is off by roughly 15%. The fact that the wider widths match better here is more an artifact of two errors countering each other. We overestimate the field strength but it is countered by an overestimate of the conductor cross-section.

> **Key Concept**
>
> A two-turn inductor with coils of the same size with minimal spacing between them and with a current moving in the same direction will double the magnetic field and quadruple the energy, hence the inductance will quadruple compared with a single-turn inductor.

An Inductor Model Including Neighboring Ground Plane

In this section we will add a ground plane underneath the inductor.

Simplify We will use the same simplifications as in the previous example for the inductor wires. For the ground plane, let us assume it is much closer to the inductor than the segments are spaced apart and that it is a perfect conductor, and that no magnetic field penetrates it. From Figure 5.27, this means $s \ll d$. From the method of images we can then see the situation mimics the previous one in that the current with the opposite sign now runs underneath in addition to horizontally. By noting that the interaction between the right quadruple and the left quadruple will tend to cancel each other when

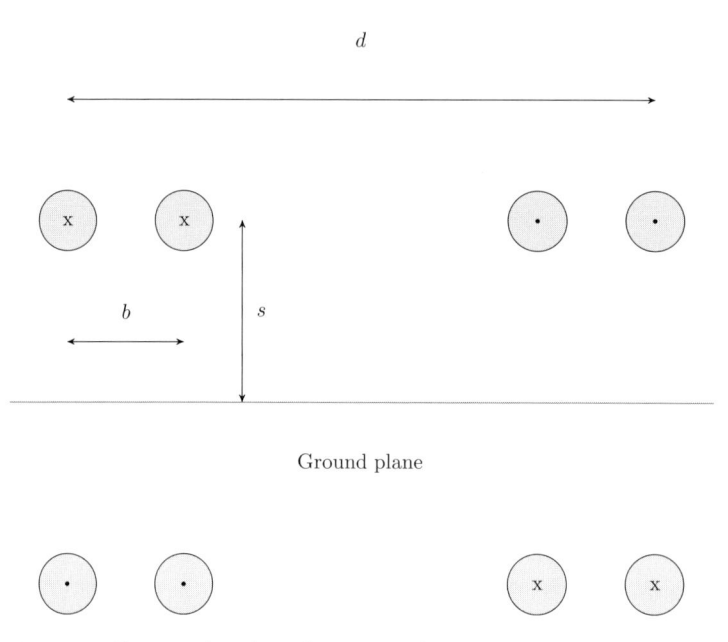

Figure 5.27 Cross-section view of a two-turn inductor over a ground plane.

they are far from each other, we can simply ignore that interaction and focus on one of them at a time.

Solve We note that we have two quadruples, so we should calculate the magnetic energy from both. However, the method of images only models the energy in the top half plane. The magnetic field in the ground plane is zero. From symmetry we can thus ignore the ground plane and one of the quadruples and calculate the total magnetic energy from one of the quadruples only. The situation is now very similar to that in the previous section. We can then simply use equation (5.14) to model the situation with d in the logarithmic term replaced by $2s$. In fact, we can simplify further by simply multiplying equation (5.14) by

$$\frac{1}{\ln(d/2s)}$$

This will work to a high degree of accuracy for the situation where $d \gg s \gg b > R_0$.

We find

$$L_{\text{2-turn, gnd}} = \frac{1}{\ln(d/2s)} \frac{2\mu d}{\pi} \left(\frac{1}{4} + \ln \frac{d}{R_0} + \frac{d-2b}{d} \left(\frac{1}{4} - 2\ln \frac{b}{R_0} + 3\ln \frac{d}{R_0} \right) \right) \quad (5.16)$$

where we have also used the fact that both turns have the same distance to the ground plane.

Verify In the following plot we show a field solver solution of the inductor structure in Figure 5.28 as a function of l with several fixed spacings and compare it with the approximate solution in equation (5.16). The simulation setup is the same as earlier, with

5.7 Inductors in Integrated Circuits

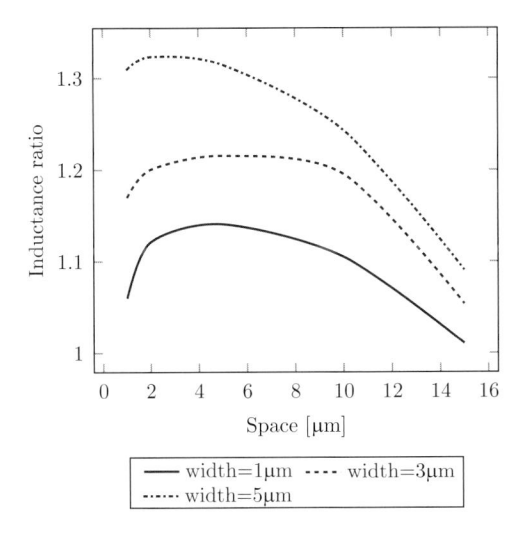

Figure 5.28 Comparison with simulation of the estimated inductance from a two-turn inductor.

the addition of a perfect ground shield underneath the inductor sitting 3 μm below the bottom of the top conductor for all cases. The field solver top conductor height was 1 μm for all cases.

Evaluate We see from the figure the wide conductor is the worst one compared to earlier. This is an artifact of the modeling here, where we use a fixed 3 μm height of the bottom of the conductor above the ground plane in the simulator. When we model this with a 5 μm-diameter cylindrical conductor in our estimation equation (5.16), it will be far from the simulator setup.

In both of these cases, 5.8.2, 5.8.4, there are a couple of things to remember.

1. The dominant effect in the inductor model is the linear term w.r.t size.
2. The less dominant effect is via the logarithmic term, which is quite forgiving error-wise.

The results shown in the figure get a little better for small spacing since the neighboring conductor is closer than the mirror image.

Overall, equation (5.16) is coarser than (5.14), and one can easily improve on it by using similar methods to those described in section "A Model of a Two-Turn Indicator." The fact that the error for the 1 μm case is quite a bit smaller than before is more related to error sources opposing each other than anything fundamentally more accurate in equation (5.16).

> **Key Concept**
>
> A ground plane underneath an inductor will reduce the inductance depending logarithmically on the distance to the inductor and the thickness/conductivity of the ground plane. The physical reason for this can be viewed as a reduction in the total magnetic field.

Summary – Two-Turn Inductor

In the previous two subsections we applied the estimation analysis to two-turn inductors. We saw that by thinking through what such an inductor looks like and comparing it with the one-turn inductor problem we already solved, we could come up with a really simple model that catches many of the aspects of the physical structure and come within 10–30% of the actual value of inductance. It is yet another good example of the benefit of estimation analysis. We will continue applying estimation analysis to the parasitic effects of real inductors such as self-resonance (due to parasitic capacitance) and resistive loss (due to resistance in the wires and substrate).

An Inductor Model Including Self-Resonance

We have looked at simplified models of the inductance of inductors. We have seen that with fairly simple assumptions we can make good predictions of inductance. Another important aspect of inductors is their self-resonance. Above this frequency it will behave as a capacitor, so it is a good thing to be able to estimate. Let us apply the method of simplification:

Simplify We showed previously in Section 5.4 that an on-chip inductor is really a high impedance transmission line that is terminated at some low impedance. The easiest way to estimate the self-resonance frequency is to calculate the $\lambda/4$ resonance as described in Section 5.4, assuming the other end of the inductor is shorted to ground.

Solve The self-resonance is simply

$$f_{res} = \frac{c}{\lambda/4} = \frac{c}{4l} \qquad (5.17)$$

where l, c is the electrical length of the inductor and c is the local speed of light. The electrical length is not always the same as the physical length since symmetries can sometimes act to reduce this length. This in turn will increase the resonant frequency. The electrical length should always be verified with a simulator. For integrated circuit use one always wants to stay away from the resonant frequency of an inductor, and if that happens to be larger due to symmetries, that is a helpful thing. The method outlined here thus gives a conservative estimate of the resonance frequency.

Verify We see in Table 5.2 our estimate of the self-resonance compared with a field solvers prediction, following the same simulation procedure we discussed earlier.
    The simulation accuracy in these simulations is around 2%.

Evaluate The self-resonance can be modeled as a short terminated transmission line. One word of caution: the electrical length can sometimes, due to boundary conditions or symmetries, be different from a naïve interpretation of just the physical length. The electrical length should always be verified with a field solver. The good news is the

5.7 Inductors in Integrated Circuits

Table 5.2 Comparison of estimated resonance frequency vs simulated

Inductor side [μm]	Resonance freq [GHz]	1-turn inductor Width 1 μm*	2-turn inductor Width 1 μm Spacing 2 μm	2-turn inductor Width 1 μm Spacing 5 μm
100	Estimated	375	95	99
	Simulated	337	94	113
	Error %	+11	+1	−12
150	Estimated	250	63	66
	Simulated	229	61	74
	Error %	+9	+3	−11
200	Estimated	187	47	49
	Simulated	174	46	55
	Error %	+7	+2	−11
300	Estimated	125	31	33
	Simulated	117	30	36
	Error %	+7	+3	−8

* The simulator finds the electrical length to be half of the physical.

electrical length is in almost all cases shorter than or equal to the physical length, so the actual self-resonance occurs at a higher frequency.

> **Key Concept**
>
> The self-resonance can be modeled as the $\lambda/4$ resonance of a short-terminated transmission line.

A Model of Inductors Including Parasitic Capacitance and Resistance

In an integrated circuit there are other conductors nearby, and a real inductor model will take them into account in various ways. We will here discuss a standard model of an on-chip inductor and how various parameters can be estimated (see [41]).

Simplify A standard model that is often used is the ad hoc or phenomenological model in Figure 5.29.

We discussed the line inductance L earlier. The capacitance can be modeled as an ideal capacitor, C_{diel}, in series with a leaky capacitor that models the silicon substrate contribution C_{sub}, R_{sub}. The series resistance R_s is simply the line resistance in the coil.

The inductor characteristics will be modeled by shorting one end to ground, so in the following one of the capacitor stacks will be grounded. This is in line with our concept of on-chip inductors as short terminated transmission lines.

Solve The capacitance C_{diel} is typically larger than the substrate capacitance, and since they are in series the substrate capacitance will dominate. We will ignore C_{diel} here.

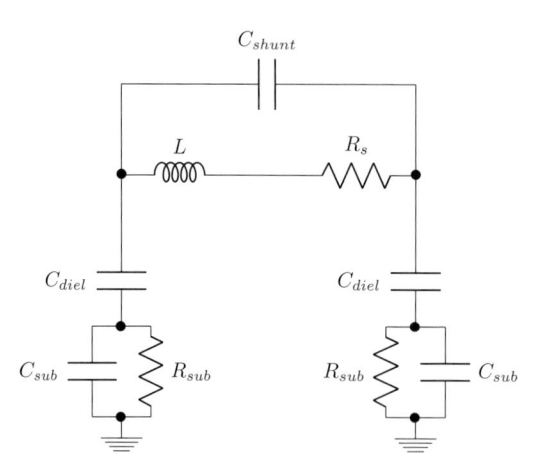

Figure 5.29 A phenomenological model of an on-chip inductor.

The shunting capacitance C_{shunt} is in parallel with the C_{sub} capacitance. We will merge them into

$$C_p = C_{shunt} + C_{sub}$$

In this simple model we see the inductor's resonance frequency is set by

$$f_{res} = \frac{1}{2\pi\sqrt{LC_p}} = \frac{c}{4l} \; \{\text{from (5.17)}\} \rightarrow$$
$$C_p = \frac{16l^2}{(2\pi)^2 c^2 L} = \frac{4l^2}{\pi^2 c^2 L} = \frac{4l^2 L/lC_t/l}{\pi^2 L} = \frac{4C_t}{\pi^2} \approx \frac{C_t}{2.5} \quad (5.18)$$

where C_t is the transmission line capacitance from the 2D calculation in Chapter 4, equation (4.55). The analysis is complicated by the fact we have two dielectric materials with different permittivity. We will only use the permittivity of the one surrounding the inductor. The substrate permittivity for bulk silicon is larger by a factor of ~4 so the effective permittivity will be larger, resulting in a smaller speed of light and a lower actual resonance frequency when compared with the estimated calculations. The fact that the final constant in (5.18) is not 2 is due to the fact that the transmission line capacitance is distributed, and in this model this capacitance has been approximated by two capacitors at each end of the inductor structure. The $1/\sqrt{LC}$ formula for the resonance frequency is valid when the capacitor shunts the full inductor. In reality, the distributed nature of the capacitance will reduce this effect and hence the factor is not 2, but rather 2.5 for this model.

The substrate resistance can be estimated by noting that the loss is due to the mostly vertical electrical field from the inductor causing currents in the substrate. If we use L_{area} to denote the area of the *conductor* portion of the inductor and T_{sub} the thickness of the substrate, we get simply

$$R_{sub} = \frac{\rho T_{sub}}{L_{area}} \qquad L_{area} = L_{length} \cdot L_{width} \quad (5.19)$$

where ρ is the substrate resistivity and L_{length} is the length of the inductor and L_{width} is its width. Often this is used as a fitting parameter when comparing with simulations/experiments, but here we will use (5.19). One can argue that for multi-turn inductors the electric field in the substrate generated by the coils will overlap if the spacing between the turns is close enough, causing an effective smaller area and higher R_{sub}. Here we will use the same formula for both single- and dual-turn inductors.

Finally, the series resistance is

$$R_s = \frac{1}{\sigma} \frac{L_{length}}{L_{thick} L_{width}} \alpha \qquad (5.20)$$

where α is a correction factor due to skin effect. This is often modeled as just a simple skin depth, δ, times circumference calculation, and in practice when comparing with simulation/measurements it is often correct within a few percent or so.

$$R_s \approx \frac{1}{\sigma \delta} \frac{L_{length}}{2(L_{width} + L_{thick})}$$

Sometimes, depending on the cross-section of the inductor, the current distribution can deviate from a simple skin depth calculation. The deviation at higher frequencies is due to effects such as the lateral skin effect discussed in Chapter 4, section "Current Distribution in Thin Conductors." Let us look at a 1×5 µm cross-section of a single-turn inductor made out of copper at 30 GHz, as shown in Figure 5.30.

Here the skin depth is ~0.4 µm, but since the height is only about 2 skin depths we still see some lingering effects of the lateral skin effect.

Another similar effect can be seen in a two-turn inductor with a width of 5 µm and a turn spacing of 1 µm displayed in Figure 5.31.

Here we also see a lateral skin effect but it is "spread out" over both conductors. This can be understood in the same way as earlier where the currents will distribute themselves to minimize the magnetic field or equivalently the inductance. In this way the magnetic near-field is a "null" between the two conductors.

In short, the field penetration into conductors is not always as simple as a skin depth calculation. Depending on conductor cross-section and frequency, the effect can be quite different. Here, unless specifically noted, we will use a simple skin depth formula.

Figure 5.30 Lateral skin effect in on-chip single-turn inductor.

Figure 5.31 Lateral skin effect in an on-chip two-turn inductor.

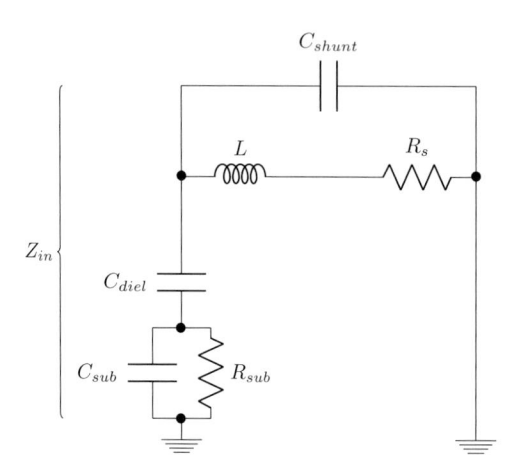

Figure 5.32 Calculation of Q from the phenomenological model.

$$\alpha = \text{MAX}\left(\frac{L_{thick}L_{width}}{2(L_{thick}+L_{width})\delta}, 1\right) \quad (5.21)$$

A key figure of merit for inductors is the quality factor, or Q. For the model shown in Figure 5.32 with one end shorted to ground we can write the impedance as

$$Z = \frac{R_{sub}/(j\omega R_{sub}C_p + 1)(R_s + j\omega L)}{R_{sub}/(j\omega R_{sub}C_p + 1) + R_s + j\omega L} = \frac{R_{sub}(R_s + j\omega L)}{R_{sub} + (j\omega R_{sub}C_p + 1)(R_s + j\omega L)}$$

$$= \frac{R_{sub}(R_s + j\omega L)}{R_{sub} + R_s - \omega^2 R_{sub}C_p L + j\omega(R_s R_{sub}C_p + L)}$$

We can rewrite this as

$$Z = \frac{R_{sub}(R_s + j\omega L)}{R_{sub} + R_s - \omega^2 R_{sub}C_p L + j\omega(R_s R_{sub}C_p + L)}$$

$$= \frac{R_{sub}(R_s + j\omega L)(R_{sub} + R_s - \omega^2 R_{sub}C_p L - j\omega(R_s R_{sub}C_p + L))}{(R_{sub} - \omega^2 R_{sub}C_p L)^2 + \omega^2(R_s R_{sub}C_p + L)^2}$$

The denominator is now real and the numerator can be written as a complex sum $a + ib$ where

$$a = R_s R_{sub}(R_{sub} + R_s - \omega^2 R_{sub}C_p L) + R_{sub}\omega^2 L(R_s R_{sub}C_p + L)$$

$$b = R_{sub}\omega L(R_{sub} + R_s - \omega^2 R_{sub}C_p L) - R_{sub}R_s\omega(R_s R_{sub}C_p + L)$$

This gives a Q when defined as an imaginary impedance divided by a real impedance

$$Q = \frac{\text{im}(Z)}{\text{re}(Z)} = \frac{R_{sub}\omega L(R_{sub} + R_s - \omega^2 R_{sub} C_p L) - R_{sub} R_s \omega (R_s R_{sub} C_p + L)}{R_s R_{sub}(R_{sub} + R_s - \omega^2 R_{sub} C_p L) + R_{sub} \omega^2 L(R_s R_{sub} C_p + L)}$$

$$= \omega \frac{L(R_{sub} - \omega^2 R_{sub} C_p L) - R_s(R_s R_{sub} C_p)}{R_s(R_{sub} + R_s) + \omega^2 L^2}$$

$$= \frac{\omega L}{R_s} R_{sub} \frac{1 - \omega^2 C_p L - R_s^2 C_p / L}{R_{sub} + R_s + \omega^2 L^2 / R_s} \qquad (5.22)$$

For low frequencies assuming $R_{sub} \gg R_s$

$$Q \sim \frac{\omega L}{R_s}$$

And for frequencies such that $\omega^2 L C_p \ll 1$ and $\omega L \gg R_s$ we find

$$Q \sim \frac{\omega L}{R_s} R_{sub} \frac{1}{\omega^2 L^2 / R_s} = \frac{1}{\omega L} R_{sub}$$

The Q quickly goes down due to substrate losses. In fact, it is the electrical field from the inductor causing currents in the substrate that is a major contributor to the loss in this case. The induced currents from the magnetic field are much smaller in size.

Verify The field solver setup is the same as discussed earlier but a lossy substrate modeled with permittivity of 11.9 and a resistivity = 0.1 ohm-m, typical of lightly doped bulk CMOS wafers, was added underneath a dielectric material with permittivity of 3 containing the inductor. The quality factor is calculated from

$$Q_{sim} = \left| \frac{\text{im } Z(1,1)}{\text{re } Z(1,1)} \right|$$

where $Z(1,1)$ is the impedance matrix response. In this case we only use one port excitation.

The top four plots in Figure 5.33 show the comparison results from a single-turn inductor with a turn width of 1 μm. It uses equation (5.22) with R_s calculated from equations (5.20), (5.21) and R_{sub} was calculated from (5.19). The bottom four plots show the results of a two-turn inductor with a width of 1 μm and a turn spacing of 1 μm.

Evaluate The quality factor of an inductor can be modeled as a loss of power in the series resistance of the inductor, and for high frequencies the substrate resistive loss needs to be included. The error in Q compared with simulations with a lumped model varies depending on size and number of turns as well as launch geometry, but varies typically between 0% and 30%. One can also see that the resonance frequency is within the same error range and estimated is mostly higher than simulated. For the capacitance evaluation in equation (5.18) the speed of light was calculated using the permittivity of the top dielectric, only resulting in too high of a value compared with the simulated value.

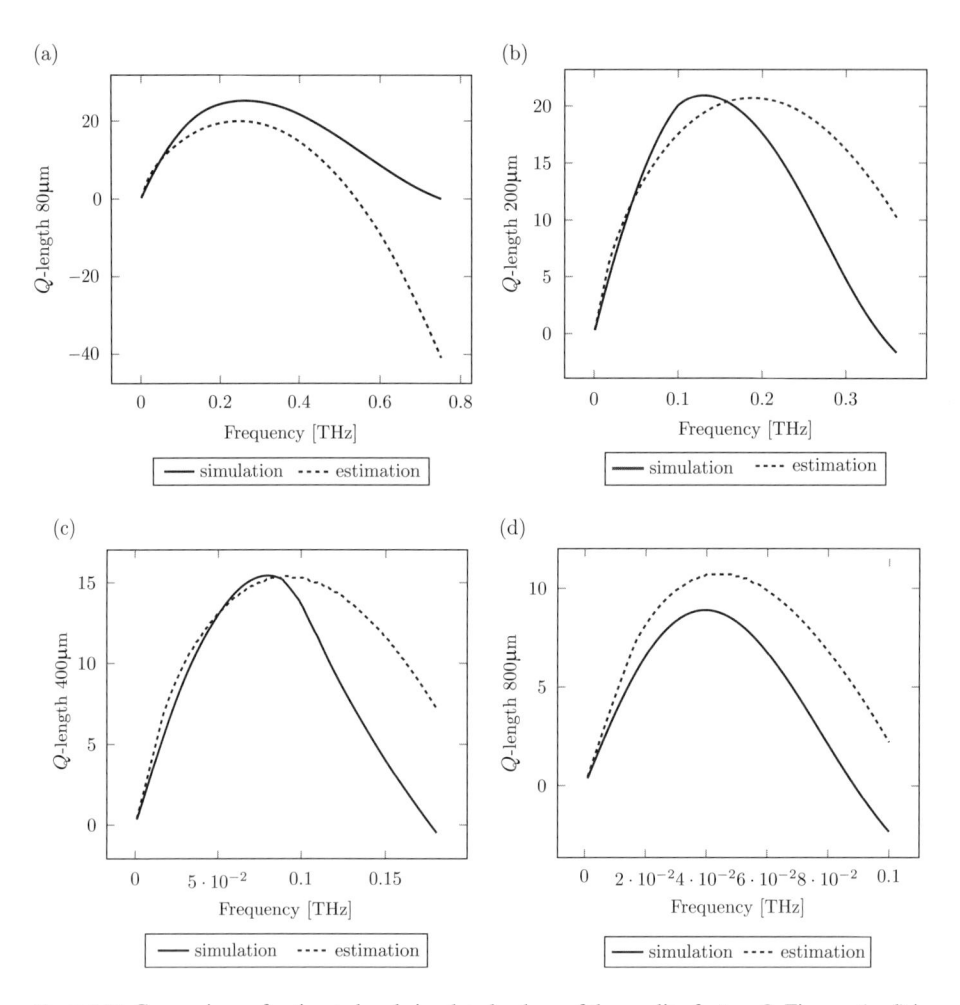

Figure 5.33 Comparison of estimated and simulated values of the quality factor, Q. Figures (a–d) is from a single turn inductor and (e–h) shows the result from a two-turn inductor.

> **Key Concept**
>
> The quality factor of an inductor can be modeled as a loss of power in the series resistance of the inductor, and for high frequencies the substrate resistive loss due to electric fields needs to be included.

5.8 Design Examples

We have investigated a number of situations and derived, using estimation analysis, a number of convenient formulae for inductors, capacitors, and transmission lines. We

5.8 Design Examples

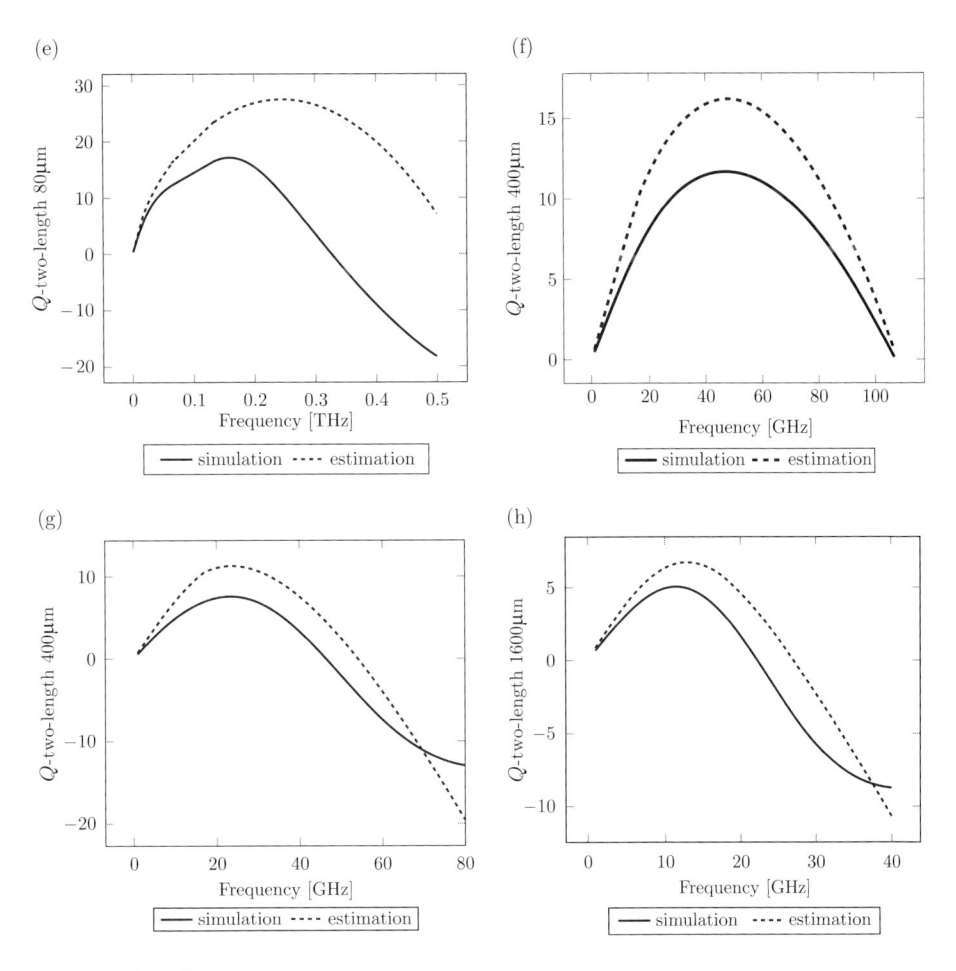

Figure 5.33 (cont.)

will here use this knowledge and construct realistic designs using these formulae as a starting point.

Example 5.1 Rectangular low-Q inductor
We will first look at a fairly large inductor to get an idea on how one can proceed. The specification in Table 5.3 is similar to what one might encounter in the "real" world.

Solution
Since 1 nH is a fairly large number, and the resulting single coil could end up being rather large, around 1 mm in circumference, let us look for a dual-turn topology. The resistance is irrelevant but we need to pay attention to the DC current. This will limit the width we can use, and since the size is expected to be small, we need to choose a minimal width that can carry the current. From Appendix A we see the maximum current density can be found for metal 9, where the limit is 5 mA/μm. This gives us the

Table 5.3 Specification table for inductor design

Specification	Limit	Comments
Inductance	1 [nH]	
Q	N/A	
DC current	5 [mA]	
Size	Small	

width, $w = 1\,\mu m$, we will use an $R_0 = 0.5\,\mu m$ following the discussion around equation (5.12) concerning equivalent area sizing.

Let us know take a guess for the length of one side. A one-turn inductor would have a size of around 200+ μm, a two-turn roughly one-quarter of this due to the coupling, which gives around 50 μm; however, the coupling is not perfect to let us shoot for $d = 75\,\mu m$ with a spacing $b = 2\,\mu m$. This gives $\ln d/R_0 = 5$. With this we can use equation (5.15) to get a good idea of a good starting point for a simulation iteration. We find

$$\frac{2\mu d}{\pi}\left(\frac{1}{2} - 2\ln\frac{b}{R_0} + 4\ln\frac{d}{R_0}\right) = 1.3\,\text{nH}$$

where we have anticipated 30% overestimation of the inductance we discussed earlier in this chapter. Plugging in the numbers, we get an updated length estimate

$$d\,8\cdot 10^{-7}\cdot 17.5 = 1.3\cdot 10^{-9} \rightarrow d \approx 92\,\mu m$$

We find also that for completion this inductor will have a resonance frequency of at least

$$f_{res} = \frac{c}{4\cdot 8\cdot l_{side}} = \frac{3\cdot 10^8/\sqrt{3}}{4\cdot 8\cdot 92\cdot 10^{-6}} = 59\,\text{GHz}$$

and a series resistance of $R_s = \rho 8 l_{side}/(w\cdot 0.5\,\mu m) = 25\,\text{ohm}$, where we have used Appendix A for some of the key parameters.

$$Q@10\,\text{GHz} \approx \frac{\omega L}{R_s} = 2.5$$

Plugging this scale into the simulator, we find after one iteration $d = 88\,\mu m$ gives $L = 986$ pH, $f_{res} = 95\,\text{GHz}$, $Q = 2.6$

Example 5.2 Rectangular high-Q inductor
Let us now look at a procedure for building a high-Q inductor. We have the specifications in Table 5.4, which, as before, can be encountered in the "real"-world.

Solution
Let us design a square inductor with a total inductance of 200 pH. We see from equation (5.10) that the inductance is more or less proportional to the length of one side. We can

Table 5.4 Specification table for 200 pH inductor design

Specification	Limit	Comments
Inductance	200 [pH]	
Q	20	@ 25 GHz
Resonance	50 [GHz]	

estimate the logarithm factor by assuming the length over width is approximately 10 giving a logarithm of 2. We then have

$$l_{side} = \frac{200 \cdot 10^{-12}}{\mu \cdot 2} \frac{\pi}{2.5} = \frac{200 \cdot 10^{-12}}{4\pi \cdot 10^{-7} \cdot 2} \frac{\pi}{2.5} \approx 100\,\mu m$$

We will verify the logarithmic term later and perhaps make some adjustments. The self-resonance for this structure is from equation (5.17) and we get

$$f_{res} = \frac{c}{4 \cdot 4 \cdot l_{side}} = \frac{1.5 \cdot 10^8}{4 \cdot 4 \cdot 100 \cdot 10^{-6}} \approx 10^{11}\ [\text{Hz}]$$

We see that at the frequency of operation we will be quite far from the self-resonance. We can then expect the dominant source of loss to be the series resistance in the inductor metal itself. We can now turn to the question of Q. We know most of the currents run along the edges of the wire due to skin effect. Let us start with a higher-level copper metal, say M9 in our process. Let us see if a 2 μm-wide wire gives us the right resistance. At 25 GHz the series resistance needs to be less than $R_s < \omega L/Q = 31/20 = 1.55$ ohm. A 400 μm-long wire of copper has a resistance $R = L\rho/A = 400\ 10^{-6}\ 1.7\ 10^{-8}/(10^{-6}\ 10^{-6} 2\ 0.5) \approx 7$ ohm, so that is too high, even when not counting the skin effect. The skin depth for copper at 20 GHz is $\delta = \sqrt{2\rho/\omega\mu} \approx 0.46\,\mu m$. It will increase the effective resistance by perhaps a factor of two. If we make the conductor wider, the skin effect will limit the reduction in resistance. It does not look like this metal layer will work. Let us try the aluminum top layer instead. We see now, $R = L\rho/A = 400 \cdot 10^{-6} \cdot 2.6 \cdot 10^{-8}/10^{-6} \cdot 10^{-6} \cdot 2 \cdot 2 = 2.6$. This looks like a better candidate. We can use a width of 4 μm to reduce the resistance more and have some margin. Having decided on the width, we can then finalize our length. We find $l_{side} = 80\,\mu m$ where we have used our knowledge that the inductance is overestimated by some 20% with this model. Our inductor should look like an 80 μm per side aluminum square with a width of 4 μm. From the formulae we find

$$L = 2 \cdot 4 \cdot 10^{-7}\ 80 \cdot 10^{-6} \left(\frac{1}{4} + \ln\left(\frac{80 \cdot 10^{-6}}{2 \cdot 10^{-6}}\right)\right) = 252\ \text{pH}$$

$$R_s = 2.6 \cdot 10^{-8} \frac{320 \cdot 10^{-6}}{0.57 \cdot 10^{-6} \cdot 2\ (4 \cdot 10^{-6} + 2 \cdot 10^{-6})} \approx 1.2$$

$$Q \approx 26\ (\text{using } L = 200\,\text{pH})$$

In summary estimation calculations give us the values in Table 5.5.

Table 5.5 Starting point size parameters for 200 pH inductor design from estimation analysis

Parameter	Value	Unit
l_{side}	80	μm
w	4	μm
L	252	pH
R	1.2	ohm
Q	26	

Table 5.6 Final size parameters after simulation optimization

Parameter	Value	Unit
l_{side}	78	μm
w	4	μm
L	200	pH
R	1.5	ohm
Q	21	

We have some margin to our design goal. We now put these parameters into a field solver and we find initially

$$L = 205\,\text{pH}$$

$$R = 1.5\,\text{ohm}$$

$$Q = 21$$

After one simulation iteration we find the final parameters in Table 5.6.

The series resistance corresponds to a shunt resistance of $R_{shunt} = Q\,\omega L \approx 660\,\text{ohm}$.

Example 5.3 **Two coupled inductors**
We will here look at the coupling effect between two square inductors and from the specifications in Table 5.7 we see we need to maximize the coupling factor.

Solution
For this example we can simply use the expression for a two-turn inductor and compare it with a single-turn inductor. As the distance between the two turns decreases, we showed earlier in section "A Model of a Two-Turn Inductor" the total inductance is close to 4× the single-turn inductor. At 4× we have a coupling factor of 1. Let us calculate the coupling factor more precisely using $b = 2R_0$. The equation for a two-turn inductor is from (5.15)

$$L_{2\text{-}turn\,b=2R_0 \ll d} = \frac{2\mu d}{\pi}\left(\frac{1}{2} + 4\ln\frac{d}{R_0} - 2\ln\frac{b}{R_0}\right).$$

5.8 Design Examples

Table 5.7 Specification table for inductor coupling

Specification	Limit	Comment
Coupling factor, k	>0.5	Assume $h/R_0 = 10$, $b/R_0 = 2$

The one-turn inductor is equation (5.11)

$$L = \frac{2\mu d}{\pi}\left(\frac{1}{4} + \ln\left(\frac{d}{R_0}\right)\right)$$

The ratio is now

$$\frac{L_{2\text{-turn}\,b=2R_0 \ll d}}{L} = \frac{2L + 2kL}{L} = 2 + 2k = \frac{\left(\frac{1}{2} + 4\ln 10 - 2\ln 2\right)}{\left(\frac{1}{4} + \ln 10\right)} = 3.2 \rightarrow k = 0.6$$

We see from this result that we need to minimize the distance between the turns as much as possible to achieve the desired coupling factor.

Example 5.4 Increase inductance by coupling

We will now use our coupling coefficient calculation in Example 5.3 to make an inductor with much larger total inductance in the same area as a single one

Solution

We can use the convenient coupling between two inductors to achieve a max of 4× the single wire inductance as in the previous example. Let us try by using a metal level M9, M10 inductors on top of each other. From Appendix A the distance between the layers is 0.5 μm. We see from Example 5.3 that with the same size inductor as in Example 5.2, we should be able to achieve close to about 3.2 times the individual inductance using the result from Example 5.3.

$$L = 3.2 \cdot 200\,\text{pH} = 640\,\text{pH}.$$

However, this is only true if the currents in the overlapping segments flow in the same direction, so great care is necessary when designing the loops. If the currents are going in the opposite direction, the designer will be very disappointed.

Example 5.5 Design an LC tank

In the previous examples we built a few inductors. We will now take the inductor in Example 5.2 and create an LC tank by adding a capacitor. We will need a resonance frequency of 25 GHz and an effective parallel resistor >500 ohm. This LC tank will be used in Chapter 7.

Solution

We can try to use the inductor we constructed in Example 5.2 and design a capacitor with a capacitance equal to

$$C = \frac{1}{L\omega^2} = \frac{1}{200 \cdot 10^{-12} \left(2\pi \cdot 25 \cdot 10^9\right)^2} = 0.2\,\text{pF}$$

We have now an equivalent circuit looking like Figure 7.6, with an equivalent parallel resistor of

$$R \approx \omega L\, Q = 660\,\text{ohm}$$

Example 5.6 Estimate capacitive load of various amplifiers

Finally, we can use what we have learned about capacitance in the first part of the chapter to estimate the parasitic capacitance of our in-design amplifiers.

Parasitic capacitance of Example 2.2

The follower amplifier designed in Example 2.2 is routed in M9 to connect to the next stage. The routing will have matched length set by the stage furthest away. We have a length of 50 μm of M9. It does not need to be wide, so we can keep it at 0.5 μm minimum width. The complication comes from the power grid routed on M10 right above M9. We find

$$C_{par} = \varepsilon \frac{A}{d} = 3 \cdot 8.85 \cdot 10^{-12} \frac{50 \cdot 10^{-6} \cdot 0.5 \cdot 10^{-6}}{0.5 \cdot 10^{-6}} \approx 1.35\,\text{fF}.$$

This calculation ignores the sidewall capacitance so we are probably off a factor of two or so. It is a about 20% of the capacitance of the next stage which is around 12 fF.

Parasitic Capacitance of Example 3.1

The comparator circuit designed in Example 3.1 has an output routing that is much smaller in length, only 5 μm, but in layer metal 2 with a width of 0.2 μm. We get

$$C_{par} = \varepsilon \frac{A}{d} = 3 \cdot 8.85 \cdot 10^{-12} \frac{5 \cdot 10^{-6} \cdot 0.2 \cdot 10^{-6}}{3 \cdot 10^{-7}} = 0.13\,\text{fF}.$$

This capacitance is very small compared with the output load of the comparator, even if we are off by a factor of two in the estimated capacitance.

Parasitic Capacitance of Example 3.2

Finally, the amplifier in Example 3.2 has an output routing in metal 8 that is 3 μm long with minimum width of 0.5 μm

$$C_{par} = \varepsilon \frac{A}{d} = 3 \cdot 8.85 \cdot 10^{-12} \frac{3 \cdot 10^{-6} \cdot 0.5 \cdot 10^{-6}}{2 \cdot 10^{-6}} = 0.02\,\text{fF}.$$

The higher-level metal routing is here very helpful in reducing the parasitic capacitance.

Example 5.7 Parasitic capacitance estimation

In this example we will estimate the parasitic capacitance of an M2 grid with length/width/spacing = 20/0.2/0.2 µm. Occasionally grids like this one appears in the daily work and a short-hand way of estimating the capacitance to ground can be useful.

Solution

The grid here is very tight, in fact from the fields perspective it will look like a fairly uniform metal. We can simply estimate the capacitance to ground assuming a solid metal as

$$C_{par} = \epsilon \frac{A}{d} = 3 \cdot 8.85 \cdot 10^{-12} \frac{20 \cdot 10^{-6} \cdot 10 \cdot 10^{-6}}{3 \cdot 10^{-7}} = 18\,\text{fF}.$$

5.9 Summary

In this chapter we have learned

- From a transmission lines analysis that an inductor is simply a transmission line terminated with an impedance much smaller than the characteristic impedance of the transmission line. Likewise, a capacitor can be seen as a transmission line that is open compared with the characteristic impedance.
- To estimate L, Q of on-die one-turn and two-turn inductors with given process parameters.
- General methodologies to approach any kind of inductor topologies.
- To estimate C of various on-die capacitors.
- To estimate S_{11}, S_{21} in various situations.

5.10 Exercises

1. Derive the inductance expressions for a rectangular cross-section with a 3-to-1 ratio. Make appropriate approximations, such as cylindrical elements, in section "An Inductor Model with Rectangular Cross-Section."
2. Estimate the size of a rectangular inductor with inductance = 1 nH, with a maximum height of 100 µm.
3. Estimate the size of an inductor with inductance = 200 pH over a ground plane with operating frequency of 10 GHz. Assume perfect ground plane.
4. Make an inductor with inductance = 3 nH in the same area as the inductor in exercise 2.
5. Make an LC tank with a resonant frequency of 10 GHz, with the effective parallel resistance >1000 ohms; is it possible? Assume a perfect capacitor.
6. A square inductor has been designed at metal M10 with a side length of 100 µm. The chip is intended to be a flip-chip and the package impact on the inductor has

not been estimated. The package can be modeled as a perfect ground plane 100 μm above the inductor. Estimate the impact of the package plane on the inductance.
7. Calculate S_{21} with a lossy transmission line and lossy ground, where both signal conductor and ground are made of copper
8. Describe shielding effects for voltages and currents and estimate good practices. Use the 1D model examples outlined in Chapter 4, sections "Simple Two-Plate System Calculation" and "First Principle Calculation of a Current Sheet over a Ground Plane" and use several thin conductor planes where the impact of the thickness of such planes on the electromagnetic fields via the vector potential A and voltage field φ can be estimated.

5.11   References

[1] H. J. Eom, *Electromagnetic Wave Theory for Boundary Value Problems*, Berlin, Germany: Springer Verlag, 2004.
[2] David M. Pozar, *Microwave Engineering*, 4th edn., Hoboken, NJ: Wiley and Sons, 2012.
[3] L. D. Landau, *Lifshitz, Electrodynamics of Continuous Media*, 2nd edn., Oxford, UK: Pergamon Press, 1984.
[4] V. Belevitch, "The Lateral Skin Effect in a Flat Conductor," *Philips Technical Review* 32, pp. 221–231, 1971.
[5] R. E. Collin, *Foundations for Microwave Engineering*, 2nd edn., IEEE Press on Electromagnetic Wave Theory, Hoboken, NJ: Wiley-IEEE, 1992.
[6] R. F. Harrington, *Time-Harmonic Electromagnetic Fields*, IEEE Press on Electromagnetic Wave Theory, Hoboken, NJ: Wiley-IEEE, 2001.
[7] Howard W. Johnson and Martin Graham, *High-Speed Digital Design: A Handbook of Black Magic*, Englewood Cliffs, NJ: Prentice-Hall, 1993.
[8] L. D. Landau and E. M. Lifshitz, *The Classical Theory of Fields*, 4th edn., Oxford, UK: Pergamon Press, 1984.
[9] A. Shadowitz, *The Electromagnetic Field*, Mineola, New York: Dover Press, 2010.
[10] B. Rojansky and V. J. Rojansky, *Electromagnetic Fields and Waves*, Mineola, NY: Dover Press, 1979.
[11] P. Hammond and J. K. Sykulski, *Engineering Electromagnetism*, Oxford, UK: Oxford University Press, 1994.
[12] J. D. Jackson, *Classical Electrodynamics*, 3rd edn., Hoboken, NJ: Wiley, 1998.
[13] C. A. Balanis, *Advanced Engineering Electromagnetics*, 2nd edn., Hoboken, NJ: Wiley, 2012.
[14] A. Niknejad and R. Meyer, *Design, Simulation and Application of Inductors and Transformers for Si RF ICs*, New York: Springer, 2000.
[15] S. Voinigescu, *High-Frequency Integrated Circuits*, Cambridge, UK: Cambridge University Press, 2012.
[16] W. M. Rogers and C. Plett, *Radio-Frequency Integrated Circuits Design*, New York: Artech House, 2003.
[17] H. Darabi, *Radio Frequency Integrated Circuits and Systems*, Cambridge, UK: Cambridge University Press, 2015.

5.11 References

[18] B. Razavi, *RF Microelectronics, Englewood Cliffs*, 2nd edn., Englewood Cliffs, NJ: Prentice-Hall, 2011.

[19] R. Martins et al., "Two-Step RF IC Block Synthesis with Pre-Optimized Inductors and Full Layout Generation In-the-loop," Transactions on Computer-Aided Design of Integrated Circuits and Systems, Early Access.

[20] G. Zhang, A. Dengi, and L. R. Carley, "Automatic Synthesis of a 2.1 GHz SiGe Low Noise Amplifier," in *Proceedings IEEE Radio Frequency Integrated Circuits Symposium*. (RFIC), pp. 125–128, 2002.

[21] G. Tulunay and S. Balkir, "A Synthesis Tool for CMOS RF Low-Noise Amplifiers," *IEEE Transactions on Computer-Aided Design of Integrated Circuits and Systems*, Vol. 27, No. 5, pp. 977–982, May 2008.

[22] R. Póvoa et al., "LC-VCO Automatic Synthesis Using Multi-Objective Evolutionary Techniques," in IEEE International Symposium on Circuits and Systems (ISCAS), pp. 293–296, June 2014.

[23] L. Chen et al., "A Novel Spiral Inductor Model with a New Parameter-Extraction Approach," Proc. IEEE International Conference on Microwave and Millimeter Wave Technology, pp. 720–723, 2010.

[24] F. Passos et al., "A Wideband Lumped-Element Model for Arbitrarily Shaped Integrated Inductors," European Conference Circuit Theory Design (ECCTD), 2013.

[25] V. Vecchi et al., "A Simple and Complete Circuit Model for the Coupling Between Symmetrical Spiral Inductors in Silicon RF-ICs," IEEE Radio Frequency Integrated Circuits Symposium, pp. 479–482, 2013.

[26] A. Ghannam et al., "High-Q SU8 Based Above-IC Inductors for RF Power Devices," IEEE, Topical Meeting Silicon Monolithic Integrated Circuits RF Systems, pp. 25–28, 2011.

[27] P. Vancorenland, C. De Ranter, M. Steyaert, and G. Gielen, "Optimal RF Design Using Smart Evolutionary Algorithms," in IEEE, Proceedings Design Automation Conference, pp. 7–10, 2000.

[28] A. Nieuwoudt, T. Ragheb, and Y. Massoud, "Hierarchical Optimization Methodology for Wideband Low Noise Amplifiers," in IEEE Proceedings Asia South Pacific Design Automation Conference, Yokohama, pp. 68–73, 2007.

[29] A. Nieuwoudt, T. Ragheb, and Y. Massoud, "Narrow-Band Low Noise Amplifier Synthesis for High-Performance System-on-Chip Design," *Microelectronics Journal*, Vol. 38, No. 12, pp. 1123–1134, Dec. 2007.

[30] Y. Xu, K. Hsiung, X. Li, L. Pileggi, and S. Boyd, "Regular Analog/RF Integrated Circuits Design Using Optimization with Recourse Including Ellipsoidal Uncertainty," *IEEE Transactions on Computer-Aided Design of Integrated Circuits and Systems*, Vol. 28, No. 5, pp. 623–637, May 2009.

[31] E. Afacan and G. Dündar, "A Mixed Domain Sizing Approach for RF Circuit Synthesis," in IEEE International Symposium on Design and Diagnostics of Electronic Circuits and Systems, pp. 1–4, June 2016.

[32] B. Liu et al., "An Efficient High-Frequency Linear RF Amplifier Synthesis Method Based on Evolutionary Computation and Machine Learning Techniques," *IEEE Transactions on Computer-Aided Design of Integrated Circuits and Systems*, Vol. 31, No. 7, pp. 981–993, July 2012.

[33] C. Ranter et al., "CYCLONE: Automated Design and Layout of RF LC-Oscillators," *IEEE Transactions on Computer-Aided Design of Integrated Circuits and Systems*, Vol. 21, pp. 1161–1170, Oct. 2002.

[34] M. Ballicchia and S. Orcioni, "Design and Modeling of Optimum Quality Spiral Inductors with Regularization and Debye Approximation," *IEEE Transactions on Computer-Aided Design of Integrated Circuits and Systems*, Vol. 29, pp. 1669–1681, 2010.

[35] R. González-Echevarría et al., "Automated Generation of the Optimal Performance Trade-Offs of Integrated Inductors," *IEEE Transactions on Computer-Aided Design of Integrated Circuits and Systems*, Vol. 33, No. 8, pp. 1269–1273, August 2014.

[36] R. González-Echevarría et al., "An Automated Design Methodology of RF Circuits by Using Pareto-Optimal Fronts of EM-Simulated Inductors," *IEEE Transactions on Computer-Aided Design of Integrated Circuits and Systems*, Vol. 36, No. 1, pp. 15–26, Jan. 2017.

[37] Y. Bontzios et al., "Prospects of 3D Inductors on Through Silicon Vias Processes for 3D ICs," IEEE International Conference on VLSI System-On-Chip, pp. 90–93, 2011.

[38] K. Salah et al., "A Closed Form Expression for TSV-Based On-Chip Spiral Inductor," IEEE International Symposium on Circuits and Systems, pp. 2325–2328, 2012.

[39] G. Yahalom et al., "A Vertical Solenoid Inductor for Noise Coupling Minimization in 3D-IC," IEEE Radio Frequency International Symposium on Circuits and Systems, pp. 55–58, 2015.

[40] B. Kim and S. Cho, "Recent Advances in TSV Inductors for 3D IC Technology," IEEE Int. SoC Design Conference, pp. 29–30, 2016.

[41] S. Mohan, "The Design Modeling and Optimization of on Hip to Chip Inductor and Transformer Circuits," Ph.D. Dissertation, Stanford University, 1999.

6 Electromagnetic Field Simulators

> **Learning Objectives**
>
> - Basic simulator principles from application of estimation analysis flow to:
>   - Long wavelength approximation
>   - Full-wave approximation

6.1 Introduction

This chapter will take our experience from Chapters 4 and 5 and use some of these results to sketch how simulators work. We will not go into great detail, as it is somewhat outside the scope of the book, but from what we have learned so far we can draw some interesting conclusions about what will make a good simulator and how a poor one may be revealed. In fact, this chapter will show yet another application of estimation analysis, although the context is quite different. We will make simplifying assumptions and briefly discuss the solution and verify/evaluate the result. There are many details when implementing a real-world simulator that fall outside the scope of this book, and we provide the interested reader with avenues to explore further in the References section below.

We start with the simple case of electrostatics where we show how charge induced from a voltage can be solved by, in principle, quite simple methods. After that we investigate inductance simulations by calculating fields from current distributions. We proceed by showing some of the basic steps necessary to design a full-wave simulator, and what an excitation port is and how it is used. Finally, a brief overview of Matrix inversions is provided.

6.2 Basic Simulator Principles

All simulators follow the same basic principles: divide the subject matter (be it time or space) into smaller chunks, referred to as a grid or mesh. Assume that inside each chunk, the property of interest is not varying at all or is varying slowly, either a constant or some linear and perhaps some higher-order polynomial. Set up the governing equations with these approximations in mind and solve them (nearly always by inverting a matrix). Repeat by refining the grid until the desired accuracy has been reached. These same principles can be found in circuit simulators, Maxwell field solvers, system simulators, etc.

For our purpose, we would like to see how field solvers operate. We have seen in previous chapters that the field from current segments can be solved precisely and, as the reader has probably guessed by now, some field solvers work by dividing up the current-carrying conductors into smaller chunks where each chunk has a constant current. Similarly, if we want to know capacitance, we divide the surface of each conductor into smaller pieces where we assume voltage and charge is either constant or slowly varying in some cases.

6.3 Long Wavelength Simulators

Long wavelength simply means that the length scales of the problem are much smaller than any field wavelength, as we saw in Chapter 4. The vector potential, A, will give rise to magnetic fields and thus inductance, whereas φ will create electric field and capacitance. We will take a look at these effects in turn.

Capacitance Simulations in Three Dimensions

We know from Chapter 4 that the long-wavelength approximation means the length scales of the problem are much smaller than any wavelength of the fields. For the potential field, we have from (4.29) that

$$\Delta \varphi(x) = 0$$

outside any conductor (where the charge is zero). This is simply Laplace's equation in three dimensions. It may be tempting to attempt to solve this equation with a given set of boundary conditions. It certainly looks fairly innocuous, but the main difficulty one runs into is one of accuracy. The result of the simulation will be induced charge on the surfaces of various conductors. To calculate the charge will involve taking the second-order derivative of φ at the surface. We know from the equation itself this derivative is zero outside. In order to implement something like this, the grid needs to be really fine and it will take a long time to solve. Instead, it is more efficient to start from (4.37) to calculate the induced charge. Here, we divide the conductors into smaller shapes where we assume the charge is constant, and using (4.37) we then calculate the potential this charge distribution produces at a certain point x. We then simply go through all the conductors, calculate the voltage resulting from everything else, and end up with a matrix equation.

We will illustrate this process by making some simplifications to illustrate the main point. It is a good example of estimation analysis but the context is a little different from the rest of the book. The estimation analysis will here result in a set of pseudocodes that

6.3 Long Wavelength Simulators

$$V_i(\mathbf{r}_i) = \int_{r'} \frac{\rho_j(\mathbf{r}')}{\sqrt{|\mathbf{r}_i - \mathbf{r}'|}} d\mathbf{r}'$$

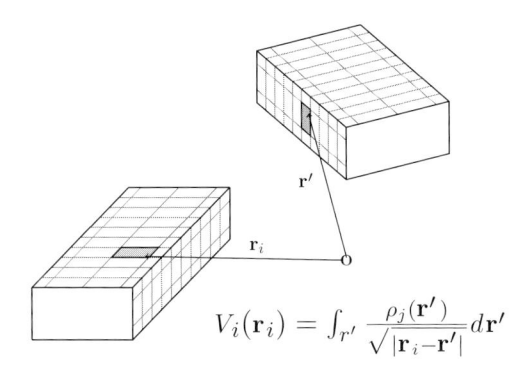

Figure 6.1 Conductors in three dimensions where the surface is divided up into smaller areas within which the charge is constant.

could be implemented. The reader must bear in mind that following these procedures will produce codes that work in principle. Building codes that are efficient and void of bugs is a whole different cup of tea and way outside the scope of the book.

Simplify We now make a number of simplifications:

- The conductors are perfect, so no charge will penetrate into the volume of the conductors, i.e., they are surface charges.
- We divide each conductor surface into much smaller surface segments where the charge is constant: see Figure 6.1.
- There is only one dielectric medium, so no dielectric boundaries need to be considered.
- We will ignore the complication of self-interaction where a given charge impacts the voltage on its own segment (there is a singularity). This is an important effect and not that difficult to deal with, but for illustrative purposes we will not address this problem here.

Solve Let us now define the problem by looking at the potential from a charge on a surface segment j, at a point $\mathbf{r}$. We have from equation (4.37)

$$V(\mathbf{r}) = \int_{Surface\_j} \frac{\rho_j(\mathbf{r}')}{|\mathbf{r} - \mathbf{r}'|} d\mathbf{r}'.$$

We now calculate this voltage at a point in the center, $\mathbf{r}_i$, of another segment i and sum over all segments. We find

$$V_i(\mathbf{r}_i) = \sum_j \int_{Surface\_j} \frac{\rho_j(\mathbf{r}')}{|\mathbf{r}_i - \mathbf{r}'|} d\mathbf{r}' = \sum_j \rho_j \int_{Surface\_j} \frac{1}{|\mathbf{r}_i - \mathbf{r}'|} d\mathbf{r}'.$$

This can be written as a matrix equation:

$$V = G \cdot \rho \tag{6.1}$$

From the boundary condition we know the value of V on all surfaces, and we can calculate G and solve for ρ. To get to capacitance, let us be a little more careful with notation. Let us assume we have N_{seg} conductor segments, each of which is subdivided into N_{sub} charge segments. We have a total number, $N_{tot} = N_{seg} N_{sub}$ of charge segments. Let us denote by ρ_{ij} charge segment j on conductor segment i. In order to find the capacitance between conductor segment i and j, we need to ground all segments except i, which has a voltage of $V_i = V$. We solve (6.1) and calculate the capacitance from (4.48)

$$C_{ij} = \frac{1}{V} \sum_{k=1}^{N_{sub}} \rho_{jk}.$$

With this procedure one needs to go through all conductors and set each voltage to V one at a time, keeping all others grounded, solve the equation, and calculate the resulting capacitance. It sounds easy enough and it sort of is, but the difficulty lies in the details. The integrand has a divergence when $i = j$ that is fairly strong and the accuracy required in the numerical integration can be excessive. Furthermore, the grid is critical and in order to set it up properly one has to have a deep understanding of how charge will distribute itself on the surfaces beforehand. However, the pseudocode is straightforward.

Pseudocode

```
Subroutine build_G(Mesh)
* Loop over Mesh to build Matrix G.
For i = 0,N do
        For j = 0,N do
                Integrate_element(r[ i] ,r[ j] )
        End for
End for
End build_G

Subroutine CapacitanceSolver
Create_mesh(Mesh)
G = Build_G(Mesh)
Ginv = Invert_matrix(G)
For i = 1,N do
        V = Define_voltages(Mesh)
        rho = Multiply(Ginv,V)
        C[ i] = sum(rho) // Assuming the voltage V = 1.
End for
End CapacitanceSolver
```

Verify Simulators are traditionally verified by having them solve known problems and comparing the solutions. We leave it as an exercise to the reader to implement a code and gain experience by learning by actually doing.

Evaluate We are not going to go to the trouble of actually building codes here, but simply state that although this may sound really simple and in some meaning of the word it is, the problem comes when you want accuracy. The charge tends to want to sit at the edges of the conductors, similar to the current distributions we examined in Chapter 5. This means that around those edges the surface elements must become really small to resolve the charge distribution, adding up to quite a few elements for large geometries. Some serious tricks need to be employed to get to the bottom of this and make the problem manageable. The most successful simulators have in common that they have found clever ways to manage the grid-generating algorithms. This has made some inventors very wealthy indeed. The construction of the grid and the speed with which the matric, G, is calculated are critical parts of building simulators. If this is done poorly it will take a long time to find a solution for a given accuracy.

Inductance Simulations in Three Dimensions

Inductance simulators can be made in a similar fashion, where we divide up the conductor into smaller current-carrying segments. The steps we need to go through are somewhat more complicated, and we will just sketch the procedure here. Far more details can be found in [3].

Simplify We assume we are dealing with one long wire of a certain thickness and width that are much smaller than the conductor length. We chop the wire into small conductor segments where we assume several things:

- The magnitude of the current in each conductor segment is the same, so there is no capacitive or other loss.
- The current is uniform in a cross-section of the wire, in other words skin depth is not important. Each conductor segment can then cover the full cross-section of the wire.
- These is only one magnetic medium, so no boundaries need to be considered.
- We will ignore the complication of self-interaction where a given current segment impacts its own magnetic field (there is a singularity). This is an important effect and not that difficult to deal with, but for illustrative purposes we do not address this problem here.

Solve Let us now define the problem by looking at the vector potential from a current on a conductor segment i, at a point r. We have from (4.36)

$$A_i(r) = \int_{Segment\_i} \frac{I_i(r')}{|r - r'|} dr'$$

With this expression we can now calculate the partial inductance matrix, $L_{i,j}$, between segments i, j, using (4.46) and

$$I_j = |I_j|e_j = I_j e_j.$$

We find

$$L_{i,j} = \frac{1}{I_i I_j} \int_{Segment\_j} I_j(r_j) \cdot A_i(r_j) dr_j = \frac{1}{I_i I_j} \int_{Segment\_j} I_j(r_j) \cdot \int_{Segment\_i} \frac{I_i(r_i)}{|r_j - r_i|} dr_i dr_j$$

$$\approx \frac{I_j \cdot I_i}{I_i I_j} \int_{Segment\_j} \int_{Segment\_i} \frac{1}{|r_j - r_i|} dr_i dr_j = e_j \cdot e_i \int_{Segment\_j} \int_{Segment\_i} \frac{1}{|r_j - r_i|} dr_i dr_j.$$

We refer to the case $i \neq j$ as the mutual inductance and $i = j$ as the self-inductance, similar to what we discussed earlier. Here we see explicitly the convenient fact that when the two currents are perpendicular to each other their mutual inductance is zero. If we decide not to divide the conductors – perhaps they are really thin, or frequency is low – we will be done here. We merely have to sum up all the elements to get the total inductance of the wire.

$$L = \sum_{i,j} L_{i,j}$$

Notice the similarities in the expressions here compared with the previous section with the capacitance solver, where we summed over all charges and divided by the voltage difference to get the total capacitance between two conductors. There are a couple of important differences. For example, for the inductance case, we assume just one wire with a certain fixed current going through all segments. This greatly simplifies the analysis and we can just calculate the coupling without having to solve a matrix equation. The normal duality of capacitance and inductance is hidden due to these different assumptions.

But we can take this one step further and look at the situation where skin effect is important. We make some additional simplifications:

- The conductors are close to ideal so the current will only flow within a certain surface depth. There is no resistive loss.
- We divide the surface of each conductor segment into much smaller surface segments within which we assume the current is constant. For simplicity, the number of subsegments is the same for all conductor segments.
- For a conductor segment, we assume the voltage drop across all surface segments is constant.

By subdividing the conductor segments into smaller segments where the current inside each is constant, we find similarly to before

$$L_{i,j} = e_i \cdot e_j \int_{Segment\_j} \int_{Segment\_i} \frac{1}{|r_j - r_i|} dr_i dr_j.$$

This defines the inductance matrix between each small surface segment. This is not quite what we want. We need the partial inductance for a particular conductor segment,

6.3 Long Wavelength Simulators

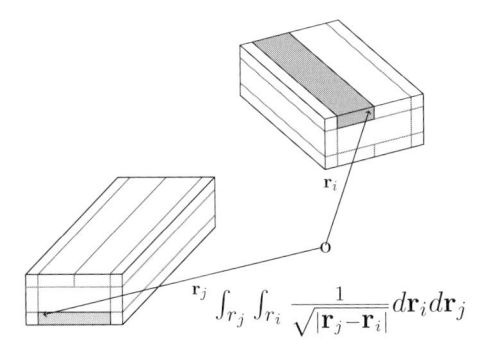

$$\int_{r_j}\int_{r_i}\frac{1}{\sqrt{|\mathbf{r}_j-\mathbf{r}_i|}}d\mathbf{r}_i d\mathbf{r}_j$$

Figure 6.2 A cross-section of a conductor with subsegments where the mutual inductance integral is indicated.

which we can then sum up to find the total inductance of the whole structure, and we do not know the size of the current in each surface segment a priori. In order to move along we need to be a little more careful with the definition of the index notation. A surface segment, i, belongs to a certain conductor k and inside this conductor segment it has a subindex, l. We then have the total number of current segments, N_{total} is the total number of conductor segments, N_{seg}, times the total number of surface-segments, N_{sub}.

$$N_{total} = N_{seg} N_{sub}.$$

A particular current segment i is then referred to as kl. Likewise for the index j we refer to a particular conductor segment m and a subsegment n (see Figure 6.2).

We now know the partial inductance for each segment, and since we also know the total current we can calculate the voltage drop across each segment assuming a particular frequency ω and using ohm's law,

$$\Delta V_{kl} = j\omega \sum_{m=1}^{N_{seg}} \sum_{n=1}^{N_{sub}} I_{mn} L_{kl,mn} = \sum_{m=1}^{N_{seg}} \sum_{n=1}^{N_{sub}} I_{mn} Z_{kl,mn},$$

with $Z_{kl,mn}$ denoting the impedance matrix. This is just a matrix equation, and we can take the inverse to find the individual currents for each surface segment

$$I_{mn} = \sum_{k=1}^{N_{seg}} \sum_{l=1}^{N_{sub}} \Delta V_{kl} Y_{mn,kl} = \sum_{k=1}^{N_{seg}} \Delta V_k \sum_{l=1}^{N_{sub}} Y_{mn,kl},$$

where $Y_{mn,kl}$ is the admittance matrix. We start to see the resemblance to the capacitance calculation in section "Capacitance Simulations in Three Dimensions." We can now take advantage of one of our simplifications, that the sum of all currents within a conductor segment is constant, conductor segment to conductor segment,

$$\sum_{n=1}^{N_{sub}} I_{mn} = I = \text{constant}.$$

We get

$$I = \sum_{n=1}^{N_{sub}} I_{mn} = \sum_{k=1}^{N_{seg}} \Delta V_k \sum_{n=1}^{N_{sub}} \sum_{l=1}^{N_{sub}} Y_{mn,kl} = \sum_{k=1}^{N_{seg}} \Delta V_k y_{m,k} \qquad (6.2)$$

The matrix $y_{m,k}$ is the admittance matrix between conductor segment m and k. After a simple inversion we find finally the impedance matrix between conductor segments.

$$z = y^{-1}.$$

And we can find the total inductance by

$$L = \frac{1}{j\omega} \sum_{i,j} z_{ij}.$$

Pseudocode

```
Subroutine build_Zij(Mesh)
* Build Matrix by looping over the Mesh
For i = 0,N do
        For j = 0,M do
                Integrate_element(r[i],r[j])
        End for
End for
End build_Aij


Subroutine InductanceSolver
Create_mesh(Mesh)
Zij = Build_Zij(Mesh)
Yij = Matrix_Inverse(Zij)
yij = Contract_subsegments(Yij)
zij = Matrix_Inverse(yij)
End InductanceSolver
```

Verify Reference [3] has a similar discussion but with more details.

Evaluate Compared with the capacitance calculation, this is somewhat more complicated, and a few more steps need to be taken if the detailed surface current distribution needs to be known. However, the same caveats we listed in section "Capacitance Simulations in Three Dimensions" apply here. Building an efficient mesh, or grid, is challenging, since the currents tend to crowd near the edges as we outlined in Chapter 5, Figures 5.30 and 5.31. Another observation is that the duality we highlighted in Chapter 4 between capacitance and inductance is nontrivial to take advantage of, in that the capacitance calculation used the fact that the values of the potential, φ, are known, and a fairly straightforward matrix equation can be set up. For the inductance case there is no boundary condition on the corresponding vector potential, A. Instead, the boundary condition is set by the total current, so a few more steps need to be taken to solve for the inductance.

6.4 Method of Moments

In the previous section we calculated the capacitance and inductance separately through their induced charges and currents. This is a great simplification and useful in the long-wavelength approximation where such concepts are well defined. In the short wavelength regime, or more popularly the full-wave regime, it is no longer possible to separate the charges and currents and we need to solve for both at once. We will briefly describe here how this can be done with the popular method of moments.

Simplify We make the following simplifications:

- The currents are constant and in a given direction within a small segment.
- There is no local charge buildup, so that all currents going out of one segment enter the next segment.
- We ignore again the problem of self-interaction due to space limitations.
- We only consider waves that are outgoing, $\sim e^{-jkr}$.

We can now divide all currents into small current segments and write the current as:

$$J(r) = \sum_{n=1}^{N} a_n I_n(r)$$

where I_n has magnitude 1 and is the direction of the local segment and a_n are the magnitudes.

Solve We get using this expression for the currents in the electric field equation (4.8) where we can write φ in terms of A via (4.12)

$$E(r) = -\mu j\omega \frac{1}{4\pi} \int \frac{J(r')}{|r-r'|} e^{-jk|r-r'|} dr' + \frac{1}{j\omega\varepsilon} \nabla \left(\nabla \cdot \frac{1}{4\pi} \int \frac{J(r')}{|r-r'|} e^{-jk|r-r'|} dr' \right)$$

$$= -\mu j\omega \frac{1}{4\pi} \int \frac{\sum_{n=1}^{N} a_n I_n(r')}{|r-r'|} e^{-jk|r-r'|} dr' + \frac{1}{j\omega\varepsilon} \nabla \left(\nabla \cdot \frac{1}{4\pi} \int \frac{\sum_{n=1}^{N} a_n I_n(r')}{|r-r'|} e^{-jk|r-r'|} dr' \right)$$

$$= -\sum_{n=1}^{N} \mu j\omega \frac{1}{4\pi} \int \frac{a_n I_n(r')}{|r-r'|} e^{-jk|r-r'|} dr' + \sum_{n=1}^{N} \frac{1}{j\omega\varepsilon} \nabla \left(\nabla \cdot \frac{1}{4\pi} \int \frac{a_n I_n(r')}{|r-r'|} e^{-jk|r-r'|} dr' \right)$$

where we have used the full-wave 3D solutions of A, φ to Maxwell's equations.

The electric field is the field resulting from the currents, commonly referred to as a scattered field, E_s. If we have an incident known field, E_i, we can apply the boundary condition that the total tangential components need to vanish on the surface of the conductor. For the part of the conductor where the incident field comes in, we have $\boldsymbol{n} \times \boldsymbol{E}_i = -\boldsymbol{n} \times \boldsymbol{E}_s$, for the rest of the conductor the tangential portion of the field, $\boldsymbol{n} \times \boldsymbol{E}_s = 0$, where $\boldsymbol{n}$ is the normal unit vector from the surface. We assign

$$\boldsymbol{E}_s(\boldsymbol{r}) = \boldsymbol{E}(\boldsymbol{r})$$

and find

$$-\boldsymbol{n} \times \boldsymbol{E}_i(\boldsymbol{r}) = \boldsymbol{n} \times \boldsymbol{E}_s(\boldsymbol{r})$$

$$= \boldsymbol{n} \times \left[-\sum_{n=1}^{N} \mu j \omega \frac{1}{4\pi} \int \frac{a_n \boldsymbol{I}_n(\boldsymbol{r}')}{|\boldsymbol{r} - \boldsymbol{r}'|} e^{-jk|\boldsymbol{r}-\boldsymbol{r}'|} d\boldsymbol{r}' \right.$$

$$\left. + \sum_{n=1}^{N} \frac{1}{j\omega\varepsilon} \nabla \left(\nabla \cdot \frac{1}{4\pi} \int \frac{a_n \boldsymbol{I}_n(\boldsymbol{r}')}{|\boldsymbol{r} - \boldsymbol{r}'|} e^{-jk|\boldsymbol{r}-\boldsymbol{r}'|} d\boldsymbol{r}' \right) \right].$$

We can consider the situation where we have an inductor, or transmission line, where the end stub has a forced incident electric field that is parallel to the conductor surface. In this way the incident field in the previous equation has a nonzero component. This equation is known as the EFIE (electric field integral equation). Similarly we can derive the MFIE (magnetic field integral equation); this equation will not contain any different information from the EFIE, but it can sometimes be of use.

This is one equation with a_n, $1 \leq n \leq N$ unknown coefficients. To solve this we can multiply by testing functions/integrals, $\int \boldsymbol{f}_m \, d\boldsymbol{r}$, whereby the boundary conditions can be evaluated at certain points using delta functions as testing functions; or, by other choices of testing functions, one can evaluate the boundary conditions over a region. One common choice is to use same testing functions as the current elements, $\boldsymbol{I}_n(\boldsymbol{r})$. This is known as Galerkin's method. In this way we integrate along the surface of the conductor where the boundary conditions are known. If we multiply the above EFIE equation with $\int \boldsymbol{I}_m \, d\boldsymbol{r}$ we get:

$$\int \boldsymbol{I}_m(\boldsymbol{r}) \cdot \boldsymbol{n} \times \boldsymbol{E}_i(\boldsymbol{r}) \, d\boldsymbol{r} = + \sum_{n=1}^{N} \mu j \omega \frac{1}{4\pi} \int \boldsymbol{I}_m(\boldsymbol{r}) \cdot \boldsymbol{n} \times \int \frac{a_n \boldsymbol{I}_n(\boldsymbol{r}')}{|\boldsymbol{r} - \boldsymbol{r}'|} e^{-jk|\boldsymbol{r}-\boldsymbol{r}'|} d\boldsymbol{r}' d\boldsymbol{r}$$

$$- \sum_{n=1}^{N} \frac{1}{j\omega\varepsilon} \int \boldsymbol{I}_m(\boldsymbol{r}) \cdot \boldsymbol{n} \times \nabla \left(\nabla \cdot \frac{1}{4\pi} \int \frac{a_n \boldsymbol{I}_n(\boldsymbol{r}')}{|\boldsymbol{r} - \boldsymbol{r}'|} e^{-jk|\boldsymbol{r}-\boldsymbol{r}'|} d\boldsymbol{r}' d\boldsymbol{r} \right).$$

This is now a matrix equation:

$$Z_{mn} a_n = b_m \qquad (6.3)$$

where $b_m = \int \boldsymbol{I}_m(\boldsymbol{r}) \cdot \boldsymbol{n} \times \boldsymbol{E}_i(\boldsymbol{r}) d\boldsymbol{r}$ integrated along the surface of the conductor, and

$$Z_{mn} = +\mu j \omega \frac{1}{4\pi} \int \boldsymbol{I}_m(\boldsymbol{r}) \cdot \boldsymbol{n} \times \int \frac{\boldsymbol{I}_n(\boldsymbol{r}')}{|\boldsymbol{r} - \boldsymbol{r}'|} e^{-jk|\boldsymbol{r}-\boldsymbol{r}'|} d\boldsymbol{r}' d\boldsymbol{r}$$

$$- \frac{1}{j\omega\varepsilon} \int \boldsymbol{I}_m(\boldsymbol{r}) \cdot \boldsymbol{n} \times \nabla \left(\nabla \cdot \frac{1}{4\pi} \int \frac{\boldsymbol{I}_n(\boldsymbol{r}')}{|\boldsymbol{r} - \boldsymbol{r}'|} e^{-jk|\boldsymbol{r}-\boldsymbol{r}'|} d\boldsymbol{r}' \right) d\boldsymbol{r}.$$

We simply invert this matrix equation and find the currents resulting from the external stimulus, $\boldsymbol{E}_i$. When the currents are known, we can use them to calculate the vector potential A and the potential φ. From these two we can now calculate the electromagnetic field at any point in 3D space.

This is a short outline of how one can write Maxwell's equations in a form that is amenable to numerical implementation. A similar discussion can be found in [2], where far more detail regarding the method and numerical implementation can be found. The convenient aspect of using the method of moments is that you only need to create a mesh where the currents could be flowing. There is no need to grid the whole three-dimensional space. The acceleration in terms of necessary computations can be quite extensive and it is nowadays a very popular simulation algorithm. Similarly to earlier, it is quite straightforward to write a code that can solve this equation, but the difficulty lies in creating efficient grids, dealing with the potential divergence in the evaluation of integrals, and building up the matrix elements efficiently.

On the Use of Ports

Ports are used to provide input stimulus to the structure. There are several ways to do this, including delta gap source ports and wave ports. We will describe them briefly but not go into implementation details.

Delta Gap Source

A delta gap source is an impressed electric field between two conductors (see Figure 6.3). This field is then used as the external field in, for example, the method of moments. It is the simplest of the ports commonly used. For the integrated circuit designer it is quite relevant because of the small size of the active devices that connect to the conductors. For microwave applications it is much less common.

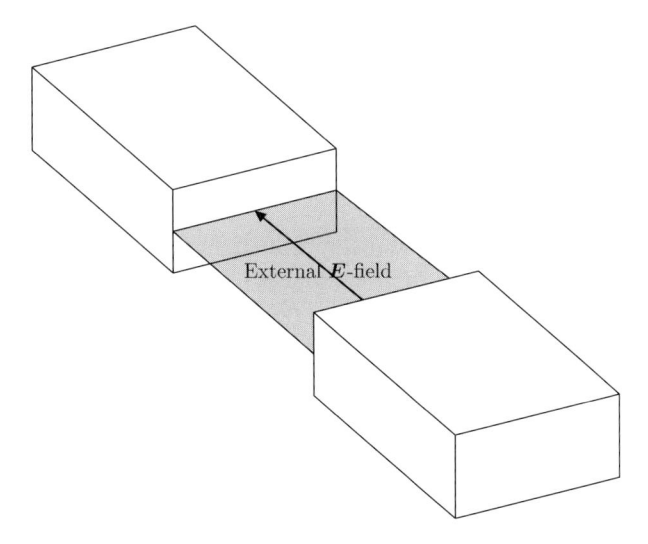

Figure 6.3 A simple 2D projection of two conductors with a field impressed between them. This is the incoming field and the way it is often pictured in the simulation setup.

Wave Port

A wave port is often defined on the boundaries of the system under consideration. Simply put, the fields are solved assuming two-dimensional symmetry on the boundary, as in Chapters 4 and 5. The intention is to mimic, say, a coaxial cable where there is an electromagnetic wave running, and it will look approximately like a field where the electric and magnetic field vectors are perpendicular to the direction of travel, hence the two-dimensional solution. Such a solution is then simply used as the boundary condition for the system under consideration. One can imagine modeling the input connector to a printed circuit board or some such application where this wave port model is a natural approximation. This type of port is very common in microwave applications.

Matrix Solvers

We have encountered matrix equations a couple of times in this chapter, and for completeness we will briefly mention how one can go about solving such equations. Basically it involves inverting the matrix and multiplying it with the right-hand side (rhs). Matrix inversion is a very active research field, and algorithms that were popular 20 years ago have been superseded by much more efficient methods. If one is interested in finding a modern implementation, an internet search is recommended. Many research groups publish their code in the public domain, so one can easily download and use it in one's own implementation, but there can be licensing limitations for commercial use. If one is interested in writing one's own code, there are broadly speaking two classes of solvers: direct matrix manipulations (exemplified by Gaussian elimination, LU decomposition) and iterative methods.

Gaussian Elimination

With Gaussian elimination one simply reorders the equations through various multiplications/divisions and row additions/subtractions to end up with a single unknown in one of the equations, which can then be solved with simple back substitution. This is then solved and the process repeats with the now reduced set of equations. For more details see [2, 6]. This generally works but requires one to redo all the work for each rhs.

LU Decomposition

Here one eliminates the problem of the rhs by writing the matrix as a product of two other matrices, L, U such that $A = LU$. The L matrix has the lower left triangle field including the diagonal, and U has the upper right triangle field with zeros in the diagonal. This way of writing the equation results in another way of doing back substitution as earlier, but it no longer depends on the rhs, and as long as the matrix is not changing it is often a better method. References [2, 6] have more details.

Iterative Methods

What have really impacted the speed of the matrix inversion algorithms are iterative methods that can be very advantageous for large systems. The basic idea is to start with a guess, x_0 to the solution of $Ax = b$. The idea is to come up with a way to minimize the residual $y = A(x - x_0)$ by somehow calculating a new solution from $x_1 = x_0 + \beta z_0$, where z_0 is some cleverly chosen direction. This goes on until the desired accuracy (size of $|y|$) has been reached. There are a number of different ways of doing this, including conjugate gradient and biconjugate gradient methods (see [2, 6]). One class of remarkably easy to implement and efficient methods that work best for sparse matrices (frequently encountered in circuit systems) are the Krylov Subspace methods. See [7] for a good discussion on these techniques.

Verify The procedure we have sketched here is similar to what is discussed in much more detail in [2].

Evaluate We arrived at the matrix equation through somewhat broad strokes, and the details of implementing such expressions in a code is outside the scope of this book. It is encouraging that the core of a simulator's operation can be understood from fairly simple arguments. It should be clear that given certain boundary conditions in terms of electric fields, we get as a result currents at various locations. It is then natural to think of the output of the simulator as Y-parameters that are defined as a current resulting from a voltage stimulus. What we are interested in as a final output is likely S-parameters, since measurement equipment is mostly set up to produce such. In [5] there is a good description of how to generate S-parameters given Y-parameters.

6.5 Summary

We have studied in a broad sense how electromagnetic field solvers can be implemented. We first looked at capacitance and inductance solvers appropriate for low frequencies, then outlined how a full wave simulator might be set up in terms of equations. Throughout the chapter we adapted the estimation analysis way of thinking. The details concerning code implementations are outside the scope of this book.

6.6 Exercises

1. Build a code that can solve for capacitances in either two or three dimensions. In particular, pay attention to the self-interacting terms. How would you avoid the potential singularity issue? For support codes such as integration algorithms and matrix solvers, there are plenty of online resources.
    a. How should the mesh (grid) be set up? Where is the charge accumulating?
2. Do likewise for inductance.

6.7 References

[1] A. Taflove and S. C. Hagness, *Computational Electrodynamics: The Finite-Difference Time-Domain Method*, 3rd edn., Norwood, MA: Artech House, 2005.

[2] W. C. Gibson, *The Method of Moments in Electromagnetics*, 2nd edn., New York: CRC Press, 2014.

[3] W. T. Weeks, L. L. Wu, M. F. McAllister, and A. Singh, "Resistive and Inductive Skin Effect in Rectangular Conductors," *IBM Journal of Research & Development*, Vol. 23, No. 6, pp. 652–660, 1979.

[4] S. M. Rao, T. K. Sarkar, and R. F. Harrington, "The Electrostatic Field of Conducting Bodies in Multiple Dielectric Media," *Transactions on Microwave Theory and Techniques*, Vol. MTT-32, No. 11, pp. 1441–1448, 1984.

[5] David M. Pozar, *Microwave Engineering*, 4th edn., Hoboken, NJ: Wiley and Sons, 2011.

[6] W. H. Press et al., *Numerical Recipes*, 3rd edn., Cambridge, UK: Cambridge University Press, 2007.

[7] Y. Saad, *Iterative Methods for Sparse Linear Systems*, 2nd edn., Philadelphia, PA: Society for Industrial and Applied Mathematics, 2003.

7 System Aspects

> **Learning Objectives**
>
> - In this chapter we will demonstrate how to apply estimation analysis to higher-level systems such as
>   - Feedback systems – PLLs
>   - Fourier transforms and how to efficiently use them when doing estimation analysis – sampling theory
>   - Defining differential equations – circuit analysis
>   - Laplace transforms – loops, both systems and circuits
>   - Simple estimates using nonlinear perturbations – VCO amplitude
>   - Sinusoidal perturbations of large signals – jitter–phase noise relationship

We assume the reader has encountered the basic mathematical theory covering the definitions of these concepts before in elementary classes. The reader does not need a background in the high-level systems themselves, as the discussion will be held at an introductory level.

7.1 Introduction

In the previous chapters we have seen a few different examples of the kind of estimation techniques that are helpful in building understanding of physical systems. In particular, the RF sections showed that a two-dimensional approximation, sometimes with additional symmetries, is of great help. There we also dug fairly deeply into certain aspects in order to more easily estimate effects such as inductance and capacitance. The character of this chapter is different. Here we will look into several physical systems where the basic approximations will be different from each other, and we will not dig into issues as deeply as we did earlier. We will paint the picture in broad strokes, with the occasional detailed analysis of systems that sometimes cause confusion to the early career engineer. Common to all analysis is a firm adherence to the belief that detailed mathematical analysis is a key to understanding systems behavior. We also show that analysis does not need to be overbearing and overly tedious. Instead, keeping the models as simple as possible without oversimplifying is the key to success. The overarching theme of the chapter is timing jitter – how it is generated, how it degrades performance, and how it can be countered.

Throughout the chapter we emphasize how to build simple yet relevant models and illustrate several different mathematical techniques. Sometimes the approximations are more or less obvious or familiar, at other times less so. We also include some well-known examples of this type of modeling from the literature to further strengthen our argument that it is both a useful and universal technique to gain understanding.

The chapter discusses phase-locked loops (PLLs) and analog-to-digital converters (ADCs). If one understands these two concepts, in effect loops and time sampling techniques, one can cover a lot of ground in the engineering space. The reader is not assumed to have any prior knowledge of either of these systems and the discussion will be kept at an introductory level for the most part. At the end of the two sections we will discuss specific design examples of certain components of the systems. We start with a discussion of clock generation in the form of PLLs and highlight the jitter aspects. Specifically we will dig into voltage-controlled oscillators (VCOs) in some detail. This is followed by a discussion of ADCs, where we design a flash-type ADC, and in particular sampling theory, where the impact of jitter and other degradations of the signal-to-noise ratio (SNR) are presented in both voltage sampling and charge sampling contexts.

7.2 Jitter and Phase Noise

Jitter

Since we will encounter the concept of jitter quite a few times in this chapter, let us first define it. Jitter is simply the deviation in time of a clock or data edge from its ideal position. In the literature one finds there are several different types of jitter discussed [1]. It is commonly divided into random jitter (RJ) and deterministic jitter (DJ). Random jitter is Gaussian in nature, with unbounded amplitude, while deterministic jitter is bounded in amplitude. The major jitter components can be further subdivided.

Deterministic Jitter

- Data-dependent jitter – this is further composed of duty cycle distortion and inter symbol interference (ISI).
- Periodic jitter – a repeating signal at a certain period or frequency.
- Bounded uncorrelated jitter – cross talk is the dominant component.

Random Jitter

- Gaussian jitter, sometimes called rms – the edge spread around the ideal arrival time is Gaussian in nature.
- Multiple Gaussian jitter – the same as above but with multiple modality.

In this chapter our focus will be on Gaussian jitter.

Phase Noise vs Jitter

This section will explore the relationship between phase noise and jitter in two different ways. In the first section we will use a simple model to get a feel for the behavior, and then we will use a more general model in the second section. In both sections we will make two simplifying assumptions, which are not strictly speaking necessary for the argument to hold but for most cases of interest they are relevant. We will assume:

1. The amplitude of the phase change is small compared with a full rotation.
2. The rate of change of the phase is small compared with the main tone.

We will quantify these assumptions in the discussion.

Simple Model

Imagine an ideal oscillator oscillating at angular frequency, ω_s. We can describe this mathematically as

$$V_{s,ideal} = A \sin \omega_s t.$$

In order to investigate the phase noise of such a system, let us start with the simple case of a single tone phase oscillation with an amplitude A_m, and frequency ω_m. We get

$$V_s = A \sin(\omega_s t + A_m \sin(\omega_m t))$$

Simplify The simplifying assumptions above mean here

1. $A_m \ll 1$
2. $\omega_m \ll \omega_s$

Solve This expression can then be expanded using simple trigonometry:

$$V_s = A \left(\sin(\omega_s t) \cos(A_m \sin(\omega_m t)) + \cos(\omega_s t) \sin(A_m \sin(\omega_m t)) \right)$$
$$\approx A \left[\sin(\omega_s t) + \cos(\omega t) A_m \sin(\omega_m t) \right]$$

Assumption 1 is a reasonable approximation since the final timing jitter is often small compared with a signals period. We can see by looking at the last term that this phase noise tone actually creates two side bands around the main tone:

$$A \cos(\omega_s t) A_m \sin(\omega_m t) = \frac{1}{2} A A_m (\cos((\omega_s + \omega_m)t) + \cos((\omega_s - \omega_m)t)).$$

In terms of frequency spectrum we have for positive frequencies

$$V_s(\omega) \sim A \, \delta(\omega - \omega_s) + \frac{1}{2} A A_m \left(\delta(\omega - \omega_s + \omega_m) + \delta(\omega - \omega_s - \omega_m) \right).$$

Phase noise is defined as the power in a single sideband divided by the power of the main tone. We find here

$$P_m = \left(\frac{(1/2)A\,A_m}{A}\right)^2 = \left(\frac{1}{2}A_m\right)^2. \tag{7.1}$$

(Note the units are often quoted as dBc/Hz since we are comparing two powers, but one can argue the units really should be radians^2/Hz.)

We will return to this observation in a little bit. For this discussion we will assume the phase jitter matters at the zero-crossing point of V_s where the slope is positive, which will happen approximately at the points in time where $\sin(\omega_s t) = 0$, and $d\sin(\omega_s t)/dt > 0$, using the assumption A_m is small (this won't work for large phase noise). Let us annotate this ideal crossing time as $t = t_n$ where n is the zero-crossing number. At this point in time $\cos(\omega t) = 1$ to first order in $\omega(t - t_n)$ and $\sin(\omega_s t) = \omega_s(t - t_n)$ and we can simplify the expression for V_s to be

$$V_s = A\,\omega\,(t - t_n) + A\,A_m \sin(\omega_m t) + O(\omega(t - t_n))^2.$$

The actual zero crossing will happen at:

$$V_s = A\,\omega\,(t - t_n) + A\,A_m \sin(\omega_m t) = 0$$

or

$$A\,\omega\,(t - t_n) = -A\,A_m \sin(\omega_m t)$$

$$(t - t_n) = -\frac{A_m}{\omega}\sin(\omega_m t) \approx -\frac{A_m}{\omega}\sin(\omega_m t_n).$$

where the last step assumes the modulation frequency $\omega_m \ll \omega$, assumption 2 above. This is a reasonable assumption since most of the time the majority of the noise comes from close to the main tone. The zero crossing is adjusted with a sinusoidal term that differs with crossing number, n. If we keep statistics of all these zero crossings we clearly see that this adjustment, or jitter, is just a sinusoid with an rms value of

$$j_t = \left\langle -\frac{A_m}{\omega}\sin(\omega_m t_n)\right\rangle = \sqrt{\frac{1}{T}\int_0^T \left(\frac{A_m}{\omega}\sin(\omega_m t_n)\right)^2 dt} = \frac{A_m}{\sqrt{2}\,\omega}.$$

If we look at this in terms of the phase noise definition in equation (7.1), we see we can also define

$$j_t = \frac{1}{\omega}\sqrt{\sum P_m} = \left\{\text{Two sidebands with power }\left(\frac{1}{2}A_m\right)^2\right\} = \frac{1}{\omega}\sqrt{2\left(\frac{1}{2}A_m\right)^2} = \frac{A_m}{\sqrt{2}\,\omega}.$$

This is simply a consequence of Parseval's theorem for this simple model, and we will look at the more general case in the next section. Here we can infer:

7.2 Jitter and Phase Noise

$$j_t = \frac{1}{\omega}\sqrt{\int_{-\infty}^{\infty} P_m \, df} = \frac{1}{\omega}\sqrt{2\int_0^{\infty} P_m \, df}.$$

where the last step assumes the phase noise is symmetric around the tone.

Verify This last expression is the often quoted phase noise–jitter relationship (see [1, 13]).

A More General Model

We can look at this in a more general way also. Instead of having an explicit tone in the sidebands, we can have a more general time dependency.

$$V_s = (A + a(t)) \sin(\omega_s t + \alpha(t)).$$

We will ignore $a(t)$ since for oscillators the term is dampened out due to nonlinear, limiting effects. In a linear system $a(t)$ will not affect the zero crossings if it is $<A$. For nonlinear systems where V_s also includes higher-order terms, these higher-order terms will cause $(A + a(t))$ to induce phase noise that is dominated by A, called AM–PM noise. We will not consider these systems here. The general assumptions from the introduction lead to the following simplifications:

Simplify
1. $\alpha(t) \ll 1, \quad \forall t$
2. $\dfrac{d\alpha(t)}{dt} \ll \omega_s, \quad \forall t$

Solve We have then

$$V_s = A \sin(\omega_s t + \alpha(t)) \approx A \sin(\omega_s t) + A\alpha(t) \cos(\omega_s t).$$

Close to zero crossings, which we define similarly to the previous section where $\sin(\omega_s t) \approx \omega_s(t - t_n)$ and $\cos(\omega_s t) \approx 1$, we get

$$V_s \approx A \omega_s(t - t_n) + A \alpha(t) \tag{7.2}$$

$$\alpha(t) = \alpha(t_n) + \frac{d\alpha(t_n)}{dt}(t - t_n) + O\left((t - t_n)^2\right).$$

For V_s to be at a zero crossing we find

$$A \omega_s(t - t_n) + A \alpha(t) = 0 \rightarrow (t - t_n) = -\frac{\alpha(t)}{\omega_s} = -\frac{\alpha(t_n)}{\omega_s} - \frac{1}{\omega_s}\frac{d\alpha(t_n)}{dt}(t - t_n) \approx -\frac{\alpha(t_n)}{\omega_s}$$

where the last step is just assumption 2 above. We can look at the time average (denoted by $\langle \cdot \rangle$) of the square of this expression

$$j_{rms}^2 = \left\langle (t - t_n)^2 \right\rangle = \frac{\left\langle \alpha(t_n)^2 \right\rangle}{\omega_s^2}.$$

We have

$$\left\langle \alpha(t_n)^2 \right\rangle = \frac{1}{2N} \sum_{n=-N}^{N} \alpha(t_n)^2 \approx \frac{1}{2N2\pi/\omega_s} \int_{-\frac{N2\pi}{\omega_s}}^{\frac{N2\pi}{\omega_s}} \alpha(t)^2 \, dt \to \int_{-\infty}^{\infty} \alpha'(t)^2 \, dt, \quad N \to \infty$$

where we define $\alpha'(t)$ as $\alpha(t_n)$ in units of $\left[\text{phase}/\sqrt{\text{time}}\right]$. We can finally use Parseval's theorem and get

$$\left\langle \alpha(t_n)^2 \right\rangle = \int_{-\infty}^{\infty} \alpha'(t)^2 \, dt = \int_{-\infty}^{\infty} |\hat{\alpha}(f)|^2 \, df = \left\{ P(f) = |\hat{\alpha}(f)|^2 \right\} = \int_{-\infty}^{\infty} P(f) \, df$$

where $\hat{\alpha}(f)$ has units $\left[\text{radians}/\sqrt{\text{Hz}}\right]$ and the single sideband noise (SSB) power, $P(f)$, has units [radians^2/Hz]. We then get

$$j_{rms} = \sqrt{\left\langle (t - t_n)^2 \right\rangle} = \sqrt{\frac{\int_{-\infty}^{\infty} P(f) \, df}{\omega_s^2}} = \frac{1}{\omega_s} \sqrt{\int_{-\infty}^{\infty} P(f) \, df} = \frac{1}{\omega_s} \sqrt{2 \int_{0}^{\infty} P(f) \, df}.$$

Verify This is the often quoted expression for jitter in terms of SSB phase noise power [1, 13].

Evaluate There is one particularly interesting thing to note about the derivation. The last term in (7.2) has units of voltage. Let us replace this with a voltage noise term $v_n(t)$ instead, so we find

$$V_s \approx A\,\omega_s(t - t_n) + v_n(t).$$

As before, for this to be at a zero crossing we find

$$A\,\omega_s(t - t_n) + v_n(t) = 0 \to (t - t_n) = -\frac{v_n(t_n)}{A\,\omega_s}$$

where we have assumed the terms v_n vary slowly compared with the main tone. We get

$$j_{rms}^2 = \left\langle (t - t_n)^2 \right\rangle = \frac{\left\langle v_n(t_n)^2 \right\rangle}{(A\,\omega_s)^2}.$$

With this we see there is no way to distinguish the added voltage noise from the phase noise. We can look at the jitter phenomenon either as a phase noise or as an added voltage noise source. This is the root cause of the somewhat confusing units, as described earlier; we can view this spectrum as either a phase or a voltage for small phase deviations. Likewise, in modern simulators we can choose to calculate jitter from a phase noise integration or by looking at the voltage noise at the zero crossings. The two approaches should obviously agree. Note that this is, strictly speaking, only true when assumption 2 is valid. For large phase deviations we do not have correspondingly large voltage noise. Instead, the voltage noise has a natural limitation referred to as linewidth.

Finally, we note that when using the voltage noise domain approach, the voltage noise source transfers to jitter as

$$j_{rms} = \sigma = \frac{\sqrt{\langle v_n(t)^2 \rangle}}{A\omega_s}.$$

The denominator $A\omega_s = dV_s/dt$. We find

$$\sigma = \frac{\sqrt{\langle v_n(t)^2 \rangle}}{dV_s/dt}. \tag{7.3}$$

This is the well-known "ohm's" law of jitter [1, 13].

Summary

The jitter–phase noise relationship is a simple calculation of the phase noise power in a sideband divided by the fundamental tone cycle frequency. A simple sinusoidal noise source is helpful when explaining the jitter–phase noise relationship.

7.3 Phase-Locked Loops

Phase-locked loops (PLLs) and various varieties of them are in common use in the semiconductor industry. These systems are discussed in many books, including [2–5]. See, in particular, [4] for an interesting nonlinear analysis. They are the key element when it comes to clock and timing generation, and many topologies are used to meet the varying key specifications that are needed. How to simplify these systems to study them analytically is generally known, and we will just present the basic theory here. The first sections describes architectures, performance criteria and common PLL sub-blocks. We then describe the general transfer function for a second-order, type 2 PLL, while the following sections describe detailed calculations of stability and noise transfer for a PLL loop. Finally, a design example is provided.

Architectures

Traditionally phase-locked loops have been divided into integer-N and fractional-N PLLs. The integer-N PLL features a simple integer divider, while the fractional-N PLL has some kind of averaging technique implemented in the divider that makes it possible for the loop to lock to a continuous range of frequencies.

Exciting PLLs have been invented recently, such as subsampling PLLs that circumvent some of the shortcomings of the established topologies.

Performance Criteria

DC Specifications

Power consumption: A key DC specification is the power consumption. In modern life, battery powered devices are very popular and keeping the power consumption low for all the circuit components is critical for market success.

AC Specifications

We describe briefly the most common AC specifications for PLLs here.

> *Loop bandwidth*: the closed loop bandwidth is a key to stability considerations. If it is too wide, the phase detector discrete sampling operation will cause problems. Depending on the cleanness of the on-chip VCO vs the reference oscillator, one might choose wide vs narrow bandwidths.
> *Phase margin*: key to loop filter design and stability.
> *Lock time*: the time it takes for a PLL to lock.
> *Jitter*: the accuracy of the resulting clock is important for ADC applications, as will be discussed later in the chapter.
> *Spur level*: spurs from the reference clock can show up in unexpected places.

PLL Sub-Blocks

A traditional PLL consists of four basic blocks: a phase detector, a VCO, a frequency divider, and a filter: see Figure 7.1. In this subsection we briefly describe them and derive some simple scaling rules.

Phase Detector

The role of the phase detector is to amplify the phase difference between two input square waves, referred to as the reference clock and the, possibly divided down, oscillator clock. It can be implemented in many different ways, and we will not go into all the possibilities here. Instead, we will take a brief look at a phase frequency detector followed by a charge pump implementation, depicted in Figure 7.2. It is a very common workhorse in the industry.

The flip-flops are usually up or down edge sensitive. Depending on which edge comes first, the upper or lower current source is turned towards the output where it sources or sinks charge from the next block, which is the filter block consisting of a

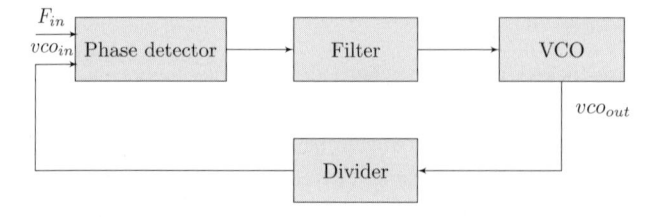

Figure 7.1 Traditional phase-locked loop topology.

7.3 Phase-Locked Loops

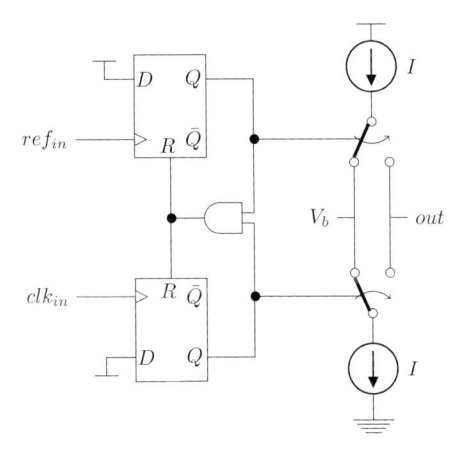

Figure 7.2 A functional view of a phase detector implemented as a phase frequency detector with a charge pump.

capacitor to ground, for simplicity. When the edge from the opposite source comes into the phase detector, the currents are turned off from the output. In this way the difference in phase between the reference and the divided down clock is translated into a current pulse. We can define the gain of this block as

$$K_{PD} = \frac{I^+ - I^-}{2\pi - (-2\pi)} = \frac{2I_c}{4\pi} = \frac{I_c}{2\pi}.$$

The phase difference can go from $+2\pi$ to -2π.

The time it will take to bring the PLL all the way from its start-up condition to phase lock is known as the pull-in time. Let us do an order of magnitude estimate of this time using estimation analysis.

Simplify First we assume the output of the phase detector sits at ground and the VCO is tuned such that the correct output frequency occurs when the control voltage at the input, which is the same as the phase detector output, is at power supply VDD, typically 7–900 mV in small geometry CMOS. The charge pump needs to bring up the output node all the way from ground to VDD. We know from our earlier discussion, on comparator analysis in Chapter 3, that this scales like

$$T_{pull\text{-}in} = \Delta t = \frac{C \Delta U}{I}.$$

The analysis is somewhat complicated by the fact the charge pump is only on for a limited time and it is possible the edge order will switch during the pull-in stage, depending on the overall dynamics. We will ignore such complications here.

Solve This is now simply a matter of plugging in the numbers, and we find

$$T_{pull\text{-}in} = \frac{C \cdot VDD}{I_c} = \frac{C \cdot \Delta \omega_{VCO}}{K_{PD} 2\pi K_{VCO}}.$$

In the last step we use the initial offset in frequency instead of voltage as a measure of how far one needs to go.

Verify This is similar to other discussions, so here we have allowed ourselves a few shortcuts to arrive at a number very similar to what others have derived (see [2]).

Evaluate It is clear that a small capacitance and a large current are helpful if a fast pull-in time is needed.

Voltage-Controlled Oscillator

The voltage-controlled oscillator will be discussed in much more detail in Section 7.4. Here we will just define the basic characteristics. The easiest way to change an oscillator's frequency is to change the effective tank load; in almost all cases this means changing a capacitance, and in most modern CMOS technologies this is natural to transistors. There are often specially constructed varactor devices with this particular property such that their capacitance is changing with its bias voltage. We can now define the gain of a voltage-controlled oscillator as K_{VCO} and we find we have an output frequency

$$\omega_{out} = K_{VCO} V_{in}$$

with a given input voltage V_{in}; note the unit of K_{VCO} is in angular frequency $2\pi f$ [MHz/V]. The K_{VCO} is assumed to be constant for estimation calculations, but in a real circuitry it will vary with voltage.

For PLL analysis we are interested in phase and not in frequency, which can often be a confusing difference. Let us think of a sine wave

$$V = \sin \omega t$$

The argument to the sinus function is a phase, but there is a frequency variable in the expression. Phase is defined as the integral of frequency. In our case in our sine wave we have a phase

$$\theta = 2\pi \int_0^t f(t') dt' = \int_0^t \omega(t') dt' = \omega t.$$

In particular, for a time varying frequency this formula can be really helpful. We can now simply relate the input voltage to the oscillator to the output phase as

$$\theta_{out}(t) = \int \omega_{out} \, dt = \int K_{VCO} V_{in} \, dt$$

which in Laplace domain corresponds to a division by s,

$$\theta_{out}(s) = \frac{K_{VCO}}{s} V_{in}(s).$$

Frequency Divider

The frequency divider simply takes an input frequency and divides it by the desired factor N. The corresponding change in phase is simply a division by N. The gain is now

$$K_{DIV} = \frac{1}{N}.$$

PLL Filter

We will discuss this block further in section "Fundamental Stability Discussion." For now, we will simply describe it as a two-port system with transfer function $F(s)$.

Basic PLL Equations

We can now pull all these block definitions together and derive the basic PLL equations. We will use the linearized transfer functions we derived in the previous section.

Simplify We will consider a phase-locked loop as in Figure 7.3. It consists of an input reference signal, a phase detector (PD), a filter, a VCO and a divider circuit. The way to simplify such a system is to linearize all the blocks and assign a gain, or transfer function, to each one. These transfer functions for the various components are illustrated in the figure. Usually, the blocks are described in terms of Laplace transformations. In a typical application one is interested in the phase transfer in the loop, and all entities here refer to phase. The variable s represents the modulation frequency around the nominal frequency. The linearization technique we describe here is known as the continuous time approximation where we ignore the fact that the phase detector–charge pump combination is actually a discrete time block. For this approximation to be valid, the loop filter bandwidth needs to be about 10× lower than the reference frequency.

Solve We start by looking at the error signal,

$$e(s) = \left(F_{in} - \frac{e(s)F(s)K_{VCO}/s}{N} \right) K_{PD}$$

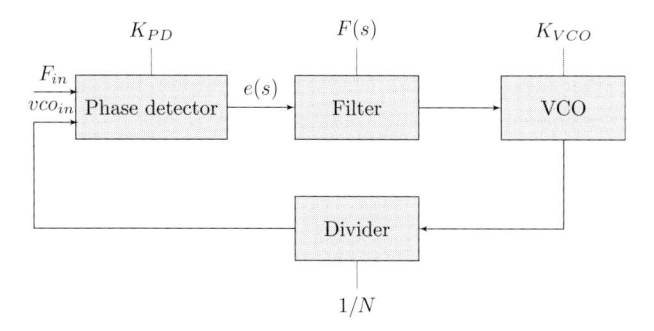

Figure 7.3 Basic PLL topology with block gains.

System Aspects

We find explicitly,

$$e(s) = \frac{F_{in}K_{PD}}{(1 + F(s)K_{VCO}K_{PD}/s\,N)}$$

We now find the transfer function $T(s) = v_o(s)/F_{in}$ from input to the VCO output

$$T(s) = \frac{e(s)F(s)K_{VCO}}{sF_{in}} \frac{K_{PD}F(s)K_{VCO}/s}{(1 + F(s)K_{VCO}K_{PD}/s\,N)} = \frac{K_{PD}F(s)K_{VCO}}{(s + F(s)\,K_{VCO}K_{PD}/N)} \quad (7.4)$$

Verify This is a well-known calculation and can be found in most textbooks on PLLs.

Evaluate Depending on the filter, we see we have at least a first-order characteristic equation in the denominator of equation (7.4).

Fundamental Stability Discussion

Stability of feedback systems is a well-studied subject, for example in [6, 7]. It comes up in many discussions, and a good understanding is very helpful in day-to-day engineering work. Here we will discuss it in the special context of PLL and second-order transfer functions. We will make some simplifying assumptions that are common in the subject and we hope the reader will be inspired to do explorations on his/her own. In the literature stability is usually discussed in terms phase margin and gain margin using Bode plots and open loop responses. This presentation uses the closed loop response to study stability. It is hoped that it will provide new insight and some variation to the more common open loop analysis. We leave it as an exercise for the reader to examine stability using open loop response.

We will use the transfer function we derived earlier and look at a couple of specific examples of the filter function, $F(s)$, and see what it implies about the system's stability.

We have for the closed loop gain

$$T(s) = \frac{K_{PD}F(s)K_{VCO}}{(s + F(s)K_{VCO}K_{PD}/N)}$$

Simplify For the filter function we will need a low-pass filter, and since the output of the charge pump is a current, the simplest low-pass filter is simply

$$F(s) = \frac{1}{sC}$$

We find then

$$T(s) = \frac{v_o(s)}{F_{in}} = \frac{K_{PD}K_{VCO}/C}{(s^2 + K_{VCO}K_{PD}/(C\,N))}.$$

7.3 Phase-Locked Loops

Solve This is a two-pole system and we can find the poles with a bit of rewrite:

$$\frac{K_{PD}K_{VCO}/C}{(s^2 + K_{VCO}K_{PD}/(C\,N))} = \frac{1}{j}\left(\frac{\sqrt{K_{VCO}K_{PD}N/C}}{s - j\sqrt{K_{VCO}K_{PD}/(C\,N)}} - \frac{\sqrt{K_{VCO}K_{PD}N/C}}{s + j\sqrt{K_{VCO}K_{PD}/(C\,N)}}\right).$$

From Appendix B we see this has the time solution

$$\sqrt{\frac{K_{VCO}K_{PD}N}{C}}\left(e^{j\sqrt{K_{VCO}K_{PD}/(C\,N)}\,t} - e^{-j\sqrt{K_{VCO}K_{PD}/(C\,N)}\,t}\right).$$

Clearly, while it does not have any increasing amplitude with time, the loop will oscillate. In most definitions of stability this situation is referred to as marginally stable, although in practice it is not acceptable. The oscillation frequency is called the natural frequency

$$\omega_n = \sqrt{\frac{K_{VCO}K_{PD}}{(C\,N)}}.$$

With this definition we find for

$$T(s) = \frac{\omega_n^2 N}{(s^2 + \omega_n^2)}.$$

The stability situation is not so good and we need to do something about that, but first let us look at the bandwidth. By replacing $s \to j\omega$ and looking at the magnitude of $T(j\omega)$, we have

$$|T(j\omega)| = \left|\frac{\omega_n^2 N}{(-\omega^2 + \omega_n^2)}\right|.$$

We see there is a singularity at the natural frequency. Let us look beyond that (denominator changes sign) and find the 3 dB bandwidth.

$$\frac{\omega_n^2 N}{(\omega_{3\,\text{dB}}^2 - \omega_n^2)} = \frac{N}{\sqrt{2}}$$

$$\omega_{3\,\text{dB}} = \omega_n\sqrt{\left(1 + \sqrt{2}\right)}.$$

In this case, the bandwidth is a tad higher than the natural frequency. As a side note, it is interesting that among the real PLLs one can buy in the market, there is often quite a bit of peaking around the natural frequency. We will see shortly that this kind of response is fairly straightforward to correct.

Verify These are all standard calculations that can be found, for example, in [2, 3].

Evaluate The expression for the bandwidth is dependent on the filter coefficients.

From a stability viewpoint, the simplification we did here in which we had a simple capacitor as integrator is simply not acceptable. In order to improve the situation we need to add a real part in the left-hand plane to the poles. We will investigate this in the next section.

Improved Stability
The suspicion is now that our loop filter is too trivial. We just have a simple integrator, a capacitor. Let us attempt a more complicated one.

Simplify

$$F(s) = \frac{1+as}{sC}$$

where $a > 0$. For low frequencies we retain our integrator action, but for high frequencies we have added a zero resulting in a constant output.

Solve Putting this into our original transfer function, we find

$$T(s) = \frac{K_{PD}F(s)K_{VCO}}{(s+F(s)K_{VCO}K_{PD}/N)} = \frac{\omega_n^2(1+as)N}{\left(s^2+(1+as)\omega_n^2\right)}.$$

Let us solve for the roots

$$s^2 + (1+as)\omega_n^2 = 0$$

$$\left(s + a\frac{\omega_n^2}{2}\right)^2 - \left(a\frac{\omega_n^2}{2}\right)^2 + \omega_n^2 = 0$$

$$s = -a\frac{\omega_n^2}{2} \pm j\sqrt{\omega_n^2 - \left(a\frac{\omega_n^2}{2}\right)^2}$$

Since $a > 0$ we see we have been successful in our quest of creating a pole in the left-hand plane. Furthermore, we see we can get rid of any oscillation by choosing

$$2\frac{1}{\omega_n} = a.$$

This particular choice is referred to as a critically damped system. Let us put this choice into the transfer function and solve for the time behavior

$$T(s) = \frac{\omega_n^2 N(1+as)}{\left(s+a\omega_n^2/2\right)^2} = \frac{2\omega_n N(\omega_n/2+s)}{(s+\omega_n)^2}.$$

We find from an inverse Laplace transform the impulse response is

$$T(t) = A\,t e^{-\omega_n t} + Be^{-\omega_n t}.$$

There is a little bit of peaking and then the exponential roll-off.

This filter is simply a resistor in series with the capacitor. The input is a current and the output a voltage. We get

$$\frac{1}{sC} + R = \frac{1}{sC}(1 + sRC)$$

and

$$RC = a = 2\frac{1}{\omega_n}$$

or

$$R = \frac{1}{C}\frac{2}{\omega_n}. \tag{7.5}$$

Finally, the bandwidth of the loop can now be estimated as

$$|T(j\omega)| = \left|\frac{2\omega_n N(\omega_n/2 + j\omega)}{(j\omega + \omega_n)^2}\right| = \frac{|T(0)|}{\sqrt{2}} = N$$

$$\frac{|T(j\omega)|^2}{N^2} = 4\omega_n^2 \frac{(\omega_n^2/4 + \omega^2)}{(\omega_n^2 - \omega^2)^2 + 4\omega^2\omega_n^2} = \frac{(\omega_n^4 + 4\omega_n^2\omega^2)}{(\omega_n^2 + \omega^2)^2} = \frac{\omega_n^4 + 4\omega_n^2\omega^2}{\omega_n^4 + 2\omega_n^2\omega^2 + \omega^4} = \frac{1}{2}$$

$$\omega_n^4 + 6\omega_n^2\omega^2 - \omega^4 = 0$$

$$\omega^2 = 3\omega_n^2 + \sqrt{10\omega_n^4} = \omega_n^2\left(3 + \sqrt{10}\right)$$

$$\omega = \omega_n\sqrt{\left(3 + \sqrt{10}\right)}.$$

Verify This is also a classic, for example, in [2–5].

Evaluate We have found a critically damped simple solution to our PLL model by making some simple assumptions and making them more complicated, to finally end up with a simple solution.

Summary

We have used estimation analysis to the full in this example and shown that we can derive some admittedly well-known results following the methodology.

> **Key Concept**
>
> A PLL's stability can be improved by inserting a zero in the filter transfer function.

PLL Noise Transfer Analysis

Having derived the basic PLL parameters, we can now investigate noise transfer. The phase noise is particularly important when calculating jitter as we discussed in section "Phase Noise vs Jitter." In this section we will discuss one noise calculation in detail and leave the rest as an exercise to the reader.

System Aspects

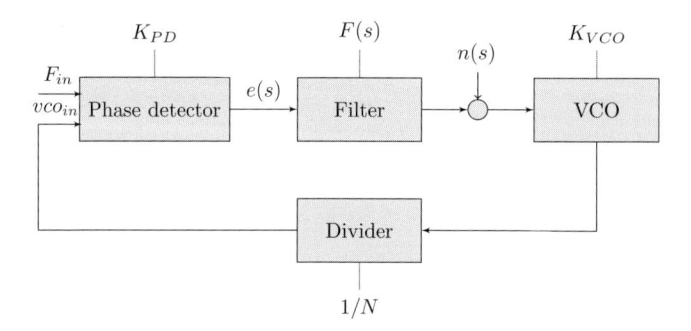

Figure 7.4 Basic PLL topology with noise injected after the filter.

Noise Injected after the Filter

Noise injected after the loop filter can be estimated by simply adding in a noise signal as shown in Figure 7.4.

Simplify We simplify the situation by considering the various blocks around their normal bias point, so we can follow the situation described earlier.

Solve Any noise injected after the filter will transfer as such:

$$(e(s)F(s) + n(s))\frac{K_{VCO}}{sN}K_{PD} = -e(s)$$

Or,

$$e(s) = \frac{-n(s)K_{VCO}K_{PD}/s\,N}{1 + F(s)K_{VCO}K_{PD}/s\,N}$$

We get at the VCO output

$$\begin{aligned}
n_F^o(s) &= \frac{(e(s)F(s) + n(s))K_{VCO}}{s} \\
&= \frac{\left(\frac{-n(s)K_{VCO}K_{PD}/s\,N}{1 + F(s)K_{VCO}K_{PD}/s\,N}F(s) + n(s)\right)K_{VCO}}{s} \\
&= \frac{n(s)K_{VCO}}{s}\left(-\frac{K_{VCO}K_{PD}}{s\,N + F(s)K_{VCO}K_{PD}}F(s) + 1\right) \\
&= \frac{n(s)K_{VCO}}{s}\left(-\frac{K_{VCO}K_{PD}}{s\,N + F(s)K_{VCO}K_{PD}}F(s) + \frac{s\,N + F(s)K_{VCO}K_{PD}}{s\,N + F(s)K_{VCO}K_{PD}}\right) \\
&= \frac{n(s)K_{VCO}}{s}\frac{s\,N}{s\,N + F(s)K_{VCO}K_{PD}} = n(s)K_{VCO}\frac{N}{s\,N + F(s)K_{VCO}K_{PD}}.
\end{aligned}$$

Verify This is a standard calculation in textbooks such as [3].

7.3 Phase-Locked Loops

Evaluate What does this mean? Let us go to various limits and observe the results, assuming $F(s) = (1 + as)/(Cs)$. For high frequencies (large s) the second term in the denominator approaches a constant. This means the loop response to the high-frequency noise is simply a low-pass filter. For low frequencies, the second term in the denominator will dominate and the noise source $n(s)$ will again be suppressed. In short, the loop acts as a band pass filter to the noise source.

Noise at the VCO Output Due to All Sources

Putting it all together, using the results from Exercise 7.1, we find at the VCO input the noise due to all noise sources in this simple model:

$$n_{VCO,output}(s) = \frac{K_{VCO}}{s} \frac{n(s)}{\left(1 + \frac{F(s)K_{VCO}K_{PD}}{sN}\right)} + \frac{K_{VCO}}{s} \frac{n_{PD}(s)F(s)}{\left(1 + \frac{F(s)K_{VCO}K_{PD}}{sN}\right)}$$

$$+ \frac{n_{VCO}(s)}{\left(1 + \frac{F(s)K_{VCO}K_{PD}}{sN}\right)} + \frac{(n_{ref}(s) - n_{div}(s))\frac{K_{VCO}}{s}K_{PD}F(s)}{\left(1 + \frac{F(s)K_{VCO}K_{PD}}{sN}\right)}.$$

We can express these noise sources in terms of the PLL loop transfer function from equation (7.4).

$$n_{VCO,output}(s) = \frac{T(s)}{K_{PD}F(s)} n(s) + \frac{T(s)}{K_d} n_{PD}(s) + \frac{n_{VCO}(s)}{N}(N - T(s))$$

$$+ (n_{ref}(s) - n_{div}(s))NT(s).$$

These noise sources are uncorrelated and the resulting noise power can be written as

$$|n_{VCO,output}(s)|^2 = \frac{1}{K_{PD}^2} \frac{|T(s)|^2}{|F(s)|^2} |n(s)|^2 + \frac{1}{K_d^2} |T(s)|^2 |n_{PD}(s)|^2 + |n_{VCO}(s)|^2 \left|\left(1 - \frac{T(s)}{N}\right)\right|^2$$

$$+ |T(s)|^2 |n_{ref}(s)|^2 N^2 + |T(s)|^2 |n_{div}(s)|^2 N^2.$$

We see clearly the high pass function of the VCO transfer where most of the remaining blocks are low pass in nature.

> **Key Concept**
>
> A PLL transfer such as the VCO noise is high pass in nature, whereas the divider reference and phase detector are low pass in nature. Depending on the filter implementation, the filter noise response is often bandpass.

Example 7.1 Block design
Previously we discussed the basic equations determining the behavior of a PLL. We will here use them to define block specs for the individual blocks in the loop. This is an

example of system design using the result of estimation analysis as a starting point. The fundamental parameters we derive can be put into a system simulator as a starting point for more detailed block-level specifications. Here, we will for brevity simply stop at the parameters provided by our simple model discussed in section "Basic PLL Equations."

PLL Specifications

Table 7.1 Specification table for PLL

Specification	Value	Comment
Output frequency	25 GHz	
Input frequency	2.5 GHz	
Output phase noise	−130 dBc/Hz	@ 1 MHz offset from tone
Bandwidth	30 MHz	

PLL Block Definitions

From the specifications in Table 7.1 it is clear we need a divide ratio of 10. Let us look at the transfer function

$$\frac{v_o(s)}{F_{in}} = \frac{e(s)F(s)K_{VCO}}{sF_{in}} = \frac{K_{PD}F(s)K_{VCO}/s}{(1 + F(s)K_{VCO}K_{PD}/(s\,N))}$$

We know $N = 10$. It remains to define K_{VCO}, C, and K_{PD}. For the charge pump we will choose a current of $I = 1$ mA and a filter capacitance of $C = 1$ pF. In a small geometry CMOS process we should have no difficulty with a 1 mA output current at 2.5 GHz. It is often better to use a higher charge pump gain than a high VCO gain due to the VCO sensitivity to the noise of the varactor. We can now look at the natural frequency and define the VCO gain.

$$\omega_n = \sqrt{\frac{K_{VCO}K_{PD}}{CN}} \leq 180 \text{ MHz}$$

Plugging in the numbers we find

$$K_{VCO} \leq \frac{3 \cdot 10^{16} \cdot 10}{10^{-3}} 2\pi \cdot 10^{-12} = 2\pi \cdot 3 \cdot 10^8 \text{ Hz/V}$$

This is a fairly reasonable number. We cover about 2.5% of the oscillator frequency and can make some adjustments for various center frequency shifts. We will use the K_{VCO} we derived here as a key specification for a VCO design later in this chapter. With a series resistor in the loop filter, it now remains to find this resistance, which we do by choosing a critically damped system. We have from equation (7.5)

7.4 Voltage Controlled Oscillators

Table 7.2 Parameters for PLL and noise spectrum

Parameter	Value	Units
N	10	
K_{PD}	$1/2\pi$	mA/rad
K_{VCO}	600	MHz/V
n_{ref}	0	V
$n(s)$	0	V
n_{PD}	$1.2 \cdot 10^{-12}$	$A/\sqrt{Hz}$
$n_{VCO}(s)$	$1/f$	$V/\sqrt{Hz}$
n_{div}	$9 \cdot 10^{-10}$	$V/\sqrt{Hz}$
C	1	pF
R	10	kohm

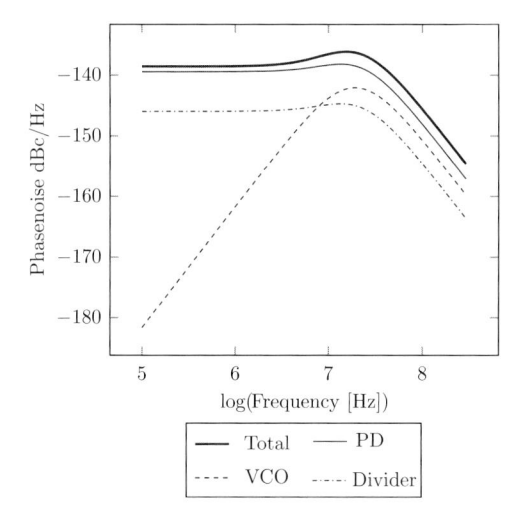

Figure 7.5 Noise sources as a function of frequency offset in PLL. Note, no 1/f noise sources are included.

$$R = \frac{2}{C\omega_n} = 10\,\text{kohm}$$

With these parameters we find the key parameters illustrated in Table 7.2.

A close-in phase noise response to the noise sources listed in Table 7.2 is shown in Figure 7.5.

7.4 Voltage Controlled Oscillators

This section describes VCOs, and following the previous discussion of PLLs, we see that voltage noise at the input will translate to phase noise at the output. At the core of almost all high-performance integrated high speed oscillators is an LC resonator that determines the frequency of oscillation and often forms part of the feedback mechanism

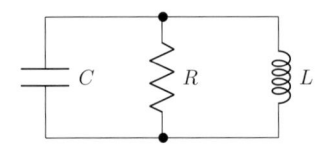

Figure 7.6 Simple model of VCO tank.

used to obtain sustained oscillations. In this section we describe how to solve for steady-state frequency, amplitude, phase noise, and finally a design example using estimation analysis. We base our discussion on references [6, 8–12].

Steady-State Frequency of Oscillation

The frequency of oscillation is naturally an important entity to understand when it comes to oscillators. We will calculate it with the help of estimation analysis.

Simplify The analysis of a high-performance oscillator begins with an analysis of a damped LC resonator, such as the parallel resonator shown in Figure 7.6.

Since there are two reactive components, this is a second-order system, which can exhibit oscillatory behavior if the losses are low or if positive feedback is added. The values are fixed only at a given frequency. All the parameters vary with frequency, where the effective parallel resistance varies the most.

Solve It is useful to find the system's response to an external stimulus. We will solve this in two ways and show they are equivalent. First we will discuss a time domain solution, which we will use later. Then we will solve the same problem in the Laplace domain. First, let us look at the circuit in Figure 7.6 and analyze it using KCL:

$$i_C + i_R + i_L = 0$$

$$\frac{di_R}{dt} R C + i_R + i_L = 0. \tag{7.6}$$

We also know the inductor's response to a change in current

$$L \frac{di_L}{dt} = u(t) = i_R R.$$

We can this use in (7.6)

$$\frac{du}{dt} C + \frac{u(t)}{R} + \int_0^t \frac{u(t')}{L} dt' = 0.$$

We now define

$$\tilde{u}(t) = \int_0^t u(t') \, dt'$$

7.4 Voltage Controlled Oscillators

and we can rewrite

$$\frac{d^2\tilde{u}(t)}{dt^2} C + \frac{d\tilde{u}(t)}{dt}\frac{1}{R} + \frac{\tilde{u}(t)}{L} = 0. \tag{7.7}$$

To solve these types of equation one can of course look up the solution in standard literature, but it is easier to work in the Laplace domain, which we will show shortly.

The general solution to this equation is

$$u(t) = A\, e^{-\frac{t}{2RC}} e^{+j\sqrt{\frac{1}{LC}-\frac{1}{4R^2C^2}}\, t} + B\, e^{-\frac{t}{2RC}} e^{-j\sqrt{\frac{1}{LC}-\frac{1}{4R^2C^2}}\, t}$$

where A, B are integration constants and can be determined from the initial conditions. The Laplace domain version of equation (7.7) becomes, by substituting $d/dt \to s$

$$s^2 u\, C + s\frac{1}{R}u + \frac{u}{L} = 0$$

$$s^2\, C + s\frac{1}{R} + \frac{1}{L} = 0$$

$$\left(s + \frac{1}{2CR}\right)^2 = \frac{1}{4\, C^2 R^2} - \frac{1}{CL}$$

$$s = -\frac{1}{2RC} \pm j\sqrt{-\frac{1}{4C^2R^2} + \frac{1}{CL}}.$$

From the solutions we see the system's response is a sinusoid with exponential decay as we found earlier. In order to maintain oscillation we need to periodically dump energy into this (tank) circuit. This is normally done by some kind of active amplifier, which is often modeled as a negative resistor in parallel with the tank.

Verify This calculation is commonplace and can be found, for example, in [8].

Steady-State Amplitude Analysis

The amplitude of oscillation is set by the point where the energy lost during one cycle is equal to the energy supplied by the active circuit. The energy lost is due to the finite quality factor of the tank circuit. It has an equivalent shunt resistor in parallel, as discussed in the previous section. The energy supplied is determined by the average of the cross-coupled negative resistor from the active circuit. Here we will discuss how to estimate the amplitude using estimation analysis.

Simplify We use as a simplification Figure 7.7. The problem can now be solved per harmonic, and here we will limit ourselves to the first harmonic only. In order to have a finite amplitude, the active circuit needs to be nonlinear to reduce the transconductance for high swing. If it does not, the amount of energy dumped into the tank will keep increasing. We will assume the active circuit current varies as such:

System Aspects

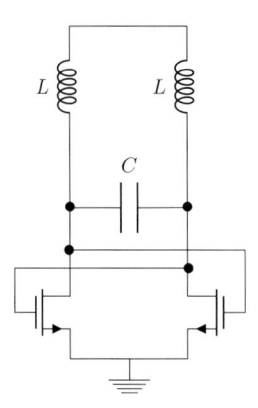

Figure 7.7 Simplified model of oscillator with tank and active (nonlinear) circuit.

$$i(v) = -g_m v + g_m'' v^3 \qquad (7.8)$$

where both $g_m, g_m'' > 0$. The negative first-order term is necessary to overcome the loss due to the real impedance in the tank, while the positive third-order term is necessary to limit the oscillation. The energy argument above means we need to find the amplitude where the absolute value of the time average admittance of the active circuit equals the time average admittance of the tank circuitry. The equation to solve is

$$|\langle Y_{active} \rangle| = \langle Y_{\tan k} \rangle$$

where $\langle \rangle$ denotes time average over one period.

Solve First we need to find the time average of the active admittance

$$y = y|_{v=0} + \left.\frac{dy}{dv}\right|_{v=0} v + \frac{1}{2}\left.\frac{d^2 y}{dv^2}\right|_{v=0} v^2 = \left.\frac{di}{dv}\right|_{v=0} v + \left.\frac{d^2 i}{dv^2}\right|_{v=0} v + \frac{1}{2}\left.\frac{d^3 i}{dv^3}\right|_{v=0} v^2$$

$$= -g_m + \frac{3}{2} g_m'' v^2$$

where we used $i(v)$ from (7.8). Expressing

$$v(t) = A \cos(\omega t)$$

we have the time averaged admittance

$$\langle Y_{active} \rangle = \frac{1}{T}\int_0^T y(t) dt = \frac{1}{T}\int_0^T \left(-g_m + \frac{3}{2} g_m'' v^2\right) dt$$

$$= \frac{1}{T}\int_0^T \left(-g_m + \frac{3}{2} g_m'' A^2 \cos^2(\omega t)\right) dt$$

We now use

$$\cos^2(\omega t) = \frac{1 + \cos(2\omega t)}{2}$$

to write

$$\langle Y_{active}\rangle = -g_m + \frac{3}{4}A^2 g_m''$$

since $\int_0^T \cos 2\omega t \, dt = 0$.

Solving for the real total admittance of the oscillator now gives us

$$-g_m + g_m'' A^2 \frac{3}{4} + \frac{1}{R} = 0$$

or

$$A = \sqrt{\left(g_m - \frac{1}{R}\right)\frac{4}{g_m'' 3}}. \tag{7.9}$$

Verify This calculation and similar discussions can be found in [8, 13].

Evaluate We see from equation (7.9) that the transconductance in the tank needs to be larger than $1/R$ of the tank, otherwise no oscillation is possible. In addition, the third-order term, $g_m'' > 0$, needs to be small for the amplitude, A, to be large, which is precisely what one might expect.

Phase Noise in Oscillators

Phase noise in oscillators has been studied in a number of papers over the years. Leeson's model has historically been very popular. It is generally considered correct, but it involves certain fitting parameters, and roughly 20 years ago [12] provided an example of a physically based model employing estimation analysis similar to what we are discussing here. The theory is linear time variant (LTV). It is time variant because the effect of phase noise depends on when during the oscillation period the noise is injected (see Figure 7.8).

The authors define what they call an impulse sensitivity function, $\Gamma(x)$, $0 \leq x \leq 2\pi$, where x is varying over the oscillation period. It simply relates the phase response to a current impulse disturbance. It will be zero at the peak of the wave form and maximum at the zero crossing. In our simple VCO model we can calculate it fairly easily, and we can then use it to find the phase noise as a function of offset frequency. For the single sideband noise, they found

$$P_{sb} = \left(\frac{i_n^2 \Gamma_{rms}^2}{q_{max}^2 (2\omega)^2}\right) \tag{7.10}$$

where Γ_{rms} is the rms-value of the impulse sensitivity function, i_n is the current noise being injected into the tank in units A^2/Hz, and ω is the angular offset frequency. We see that if the current noise is white, the frequency offset will vary such as $1/\omega^2$; if instead it varies as $1/f$, the phase noise will exhibit the well-known $1/f^3$ slope. All these parameters are fairly easily understood with the possible exception of the impulse

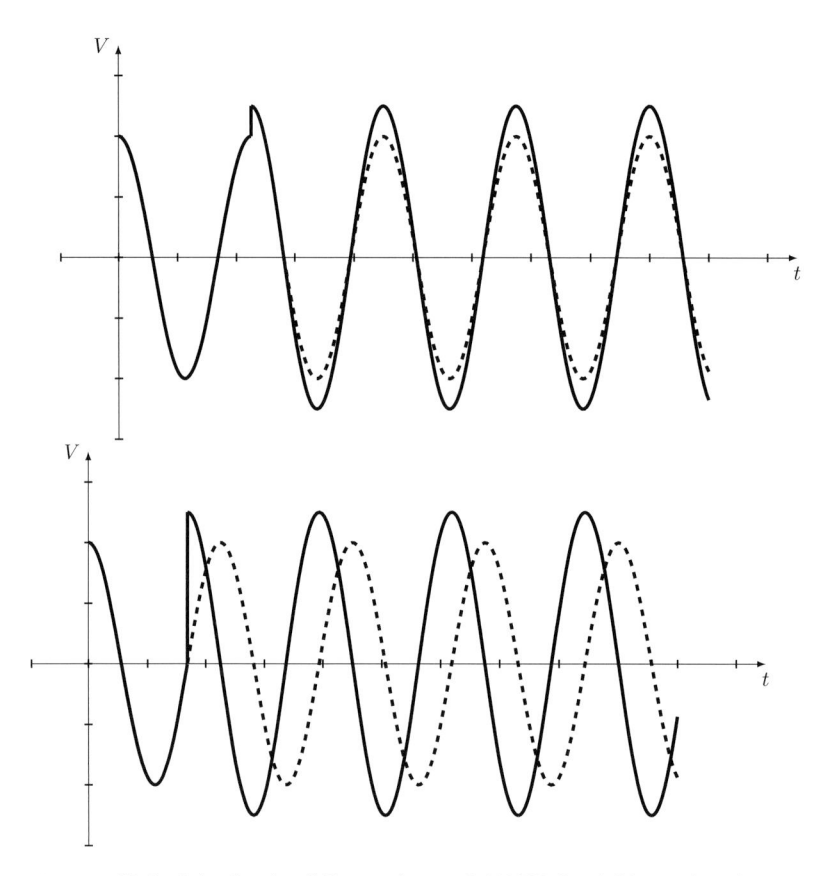

Figure 7.8 Noise injection for different phases. © [1998] Cambridge University Press. Reprinted, with permission, from Cambridge University Press.

sensitivity function. To illustrate we will calculate it here in our simplified model of the tank. We start with the basic differential equation (7.7)

$$\frac{d^2\tilde{u}(t)}{dt^2}C + \frac{d\tilde{u}(t)}{dt}\frac{1}{R} + \frac{\tilde{u}(t)}{L} = q_n\delta(t-\tau)$$

The delta function to the right is now due to an impulse current injected at time $t = \tau$.

Simplify We can now use the fact that the effective resistance is infinite due to the cross-coupled transconductor providing an opposite resistance that cancels the loss. We have

$$\frac{d^2\tilde{u}(t)}{dt^2}C + \frac{\tilde{u}(t)}{L} = i_n\delta(t-\tau)$$

Note, dimensionally $[q_n]$ = A s or ampere-second to work out properly. In other words, it is a charge.

Solve The easiest way to solve these kinds of equations is to go to the Laplace domain and solve it there. We find

$$s^2 \tilde{U}(s) + \frac{\tilde{U}(s)}{LC} = \frac{q_n}{C} e^{-s\tau}$$

This is easily solved

$$\tilde{U}(s) = \frac{q_n e^{-s\tau}/C}{s^2 + 1/LC} = -\frac{q_n}{C} \frac{\sqrt{LC}}{2j} \left(\frac{e^{-s\tau}}{s + j/\sqrt{LC}} - \frac{e^{-s\tau}}{s - j/\sqrt{LC}} \right)$$

This can be converted back to time domain and we find

$$\tilde{u}(t) = j \frac{q_n}{2} \sqrt{\frac{L}{C}} \left(e^{-j\omega(t-\tau)} - e^{j\omega(t-\tau)} \right), \quad t > \tau$$

For the voltage we now find

$$u(t) = j \frac{q_n}{2} \sqrt{\frac{L}{C}} \left(-j\omega e^{-j\omega(t-\tau)} - j\omega e^{j\omega(t-\tau)} \right) = \frac{q_n}{2} \frac{1}{C} \left(e^{-j\omega(t-\tau)} + e^{j\omega(t-\tau)} \right)$$
$$= \frac{q_n}{C} \cos(\omega(t-\tau)), \quad t > \tau$$

This is now a small disturbance added to the tank at $t > \tau$. Assume now the main oscillation is

$$v_{osc} = A \sin(\omega t)$$

If we now add the disturbance due to the injected current at time $t = \tau$ we find

$$v_{osc} = A \left(\sin(\omega t) + \frac{q_n}{AC} \cos(\omega(t - \tau)) \right)$$

The perturbation term can now be identified with the impulse sensitivity function we were looking for, which can be illustrated with a couple of cases.

Verify Let us look at specific values of τ. If $\tau = 0$ we have a simple sine wave with a zero crossing at $t = 0$

$$v_{osc} = A \left(\sin(\omega t) + \frac{q_n}{AC} \cos(\omega t) \right) \approx A \sin \left(\omega t + \frac{q_n}{CA} \right)$$

which follows from a trivial trigonometric expansion. All the injected power goes into phase noise. If instead $\tau = \pi/(2\omega)$, we find

$$v_{osc} = A \left(\sin(\omega t) + \frac{q_n}{AC} \cos \left(\omega t - \frac{\pi}{2} \right) \right) = A \sin(\omega t) \left(1 + \frac{q_n}{AC} \right).$$

The injected noise is pure amplitude noise. This is all we can expect intuitively. Remember the forced units on q_n [A s].

System Aspects

We see the perturbation term is very close to equation (7.10) and we identify

$$\Gamma(x) = \cos(x)$$

where $0 \leq x \leq 2\pi$.

Evaluate In summary, we find for the perturbation and its impact on phase noise with a charge impulse q_n as a function of phase

$$v_{perturb}(x) = \frac{q_n}{AC}\Gamma(x) = \frac{q_n}{q_{max}}\Gamma(x)$$

where

$$\Gamma(x) = \cos(x)$$

where q_{max} is a normalization constant discussed in [12].
And for the rms value,

$$\Gamma_{rms} = \frac{1}{\sqrt{2}} \qquad (7.11)$$

For LC oscillators this expression will give us a good estimate of the magnitude of the noise.

Key Concept

A VCO's jitter transfer is linear and time variant, LTV, in nature. In other words, the effect of noise on jitter depends on when the noise is injected.

Example 7.2 VCO design

We are now at a stage where a discussion of a VCO design is in order. We will build a VCO with bits and pieces we have designed in the previous chapters.

VCO Specifications

Table 7.3 shows the specifications for the VCO.

Table 7.3 Specification table for VCO

Specification	Value	Comment
Output frequency	25 GHz	
Gain	300 MHz/V	= 1800 MHz/V in angular frequency.
Output phase noise	−140 dBc/Hz	@ 30 MHz offset from tone
Supply voltage	0.9 V	

7.4 Voltage Controlled Oscillators

VCO Design

As we have discussed previously, a VCO consists mainly of some kind of resonant element, some kind of loop or a tuned resonant circuit, and an active element that replenishes the loss in this loop or resonant element. We will here use our results from the estimation analysis to:

- Get a starting point of the design work from the estimation analysis;
- Use this starting point in the simulator where we optimize the parameters.

This is a circuit example of how to use estimation analysis to speed up the design work. A good understanding of the basic behavior is essential to a good-quality end product. The simulator should not give us any surprises; it should just be used to fine-tune the performance, as we will discuss here.

We will focus on an LC VCO. We have two major design tasks. First we need to design the LC tank. We will draw from Chapter 5 to find an inductor with a specific Q. We will then shunt this with a capacitor from the same chapter to define the overall Q of this resonator at 25 GHz, which is the oscillation frequency. When we know the effective shunt loss resistor, we will be prepared to design the cross-coupled pair with a specific g_m to overcome the loss. Generally, if we are using a fast enough process, the major headaches will involve the passive tank circuitry. We also need to isolate the output of the VCO from the load of the circuitry that follows. We will therefore use the CD stage we defined in Chapter 2 at the output. The topology we now have is illustrated in Figure 7.9.

Tank Design

To resonate, we need an inductor shunted by a capacitance. We learned in Chapters 4 and 5 that the inductance is set by the current pattern as it flows in a loop. It is therefore most important to include the full current loop, *including* the flow through the active circuitry, as shown in Figure 7.10.

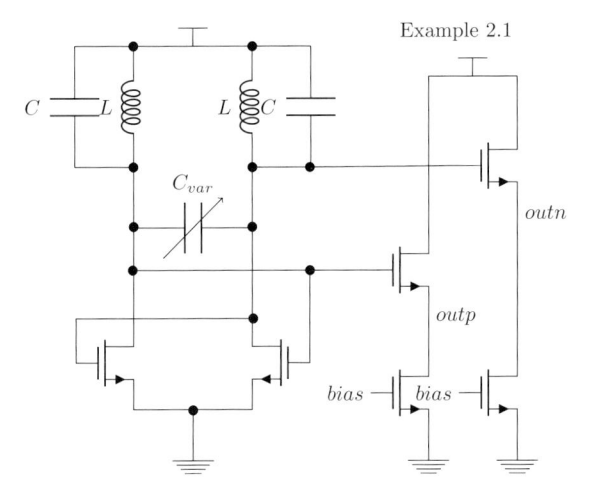

Figure 7.9 Basic VCO topology.

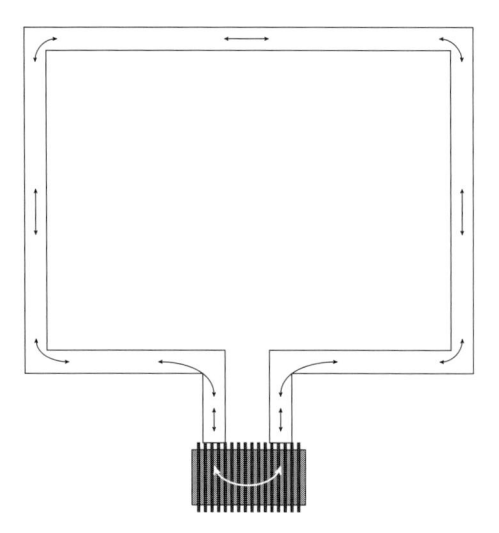

Figure 7.10 VCO inductor including active circuitry showing AC current path loop.

Do not just include the coil itself. We have also learned that the resonance frequency is set to be the length of the inductor element, so we will attempt the design by having a single-ended inductor of

$$L = 100 \text{ pH}.$$

This gives a single-ended capacitance of

$$C = \frac{1}{L\omega^2} = \frac{1}{100 \cdot 10^{-12} (2\pi \cdot 25 \cdot 10^9)^2} = 0.4 \cdot 10^{-12} \text{ F}$$

A rather large required capacitance is helpful in that the load on the tank can be included as part of the resonance. What will such an inductor look like? We created such a tank in Example 5.5, but the inductor in the tank was just the coil itself; now we need to include the connections from it to the active circuitry to get the inductance right. We will use a length of $L_{int} = 40$ μm, where the distance between these legs is $d = 10$ μm. From equation (4.52) we find the added differential inductance due to the legs is

$$L_{add} = 4 \cdot 10^{-7} \cdot 40 \cdot 10^{-6} \left(\frac{1}{4} + \ln \frac{10}{2} \right) \approx 29 \text{ pH}$$

which is overestimated by about 30%, so we are off with the inductance by about 10%. The required capacitance is then 10% lower, or $C = 0.36$ pF.

In order to adjust the frequency, a varactor needs to be included in the tank design. This device will change the capacitance and so the resonance frequency as a function of its bias voltage. From the specifications we see we need a gain of 300 MHz/V leading to a shift in frequency of 300 M/25 G = 1.2% variation, leading to a change in capacitance of 2.4%. The capacitance needs to change by 8.5 fF with one volt change. From the technology we find the varactor needs to have a size of roughly $m = 2$.

Active Design

We now know the tank specs and can proceed with the active design. As a rule of thumb, the worst case is $1/g_m < R/2$. This is to ensure we have enough margin to start the oscillations and to get a reasonable yield. We know the differential tank resistance R_{diff} from Example 5.5, from which we get the single-ended resistance R and can thus calculate the transconductance needed

$$g_m > 2\frac{1}{R} \approx 6 \text{ mmho}$$

With this oscillator design we cannot use the thin-ox transistors due to over voltage stress. We instead use a 1.5 V unit transistor from our fictitious technology. From our tabulated transistor we then get a size of 10 fingers device from Appendix A. This transistor has a transconductance $g_m \approx 6$ mmho and should be enough to get the oscillation going.

Frequency: With these parameters we see that the capacitance of the active stage can be estimated using the calculation in Exercise 2.5. We find $C_{load,act} = C_g + 4\,C_d = 21$ fF. The tank capacitance should then be reduced to $C_{tank} = 339$ fF. With the addition of the varactor, which has a nominal capacitance of 10 fF per unit, we need two plus the load CD stage, which has an input capacitance of 9 fF according to Example 2.1, so the final tank capacitor needed has an estimated size of $C_{tank} = 310$ fF.

Amplitude: We have from equation (7.9) that the amplitude will be close to

$$A = \sqrt{\left(g_m - \frac{1}{R}\right)\frac{4}{g_m''\cdot 3}} = \sqrt{\left(5.8\cdot 10^{-3} - \frac{1}{330}\right)\frac{4}{5\cdot 10^{-3}\cdot 3}} = 0.86 \text{ V}$$

Notice that the negative resistance due to the "rotated" capacitor at the output of the follower, see Example 2.1, is negligible compared with the negative resistor provided by the cross-coupled pair.

Phase noise: The expressions (7.10) and (7.11) show the phase noise to be expected is

$$\mathcal{L}(\Delta\omega) = 10\log\left(\frac{\Gamma_{rms}^2}{q_{max}^2}\cdot\frac{\overline{i_n^2}/\Delta f}{4(\Delta\omega)^2}\right)$$

where

$$\Gamma_{rms} = \frac{1}{\sqrt{2}}$$

$$q_{max} = C_{tank}A$$

$$\frac{\overline{i_n^2}}{\Delta f} = 4kTg_m\gamma.$$

Table 7.4 Starting point parameters for VCO

Device	Parameter	Value
Inductor	Inductance	110 pH
Capacitor	Capacitance	310 fF
Varactor	Multiplicity	2
M_1, M_2, medium thickness	W/L/NF	1 µm/100 nm/10

Table 7.5 Final sizes for VCO after simulation optimization

Device	Parameter	Value
Inductor	Inductance	110 pH
Capacitor	Capacitance	315.6 fF
Varactor	Multiplicity	2
M_1, M_2, medium thickness	W/L/NF	1 µm/100 nm/10

We assume here, and it is easy to verify, that the noise from the load resistor itself is small compared with the transistor noise. This is a single-ended expression. The differential impact on phase noise is 3 dB lower.

Finally, we have the estimated parameters/sizes for the components of the VCO in Table 7.4.

Simulation

After implementing these sizes in the simulator, we find that the capacitance needed is underestimated, and a slightly updated table is shown in Table 7.5.

The main error in the capacitance is that we have assumed a full $C_{load,act}$ at all times. In reality the load capacitance will vary over the cycle for all transistors.

After this has been sized, we confirm the operation with a simple simulation in Figures 7.11 and 7.12.

From the figures it is clear our correlation is very good for higher offset frequencies. For lower frequencies the $1/f$ noise of the transistors starts to come into play. Upon closer examination there are a couple of competing effects that cancel each other out. The transistor noise is overestimated, but it is compensated by the fact we do not include the other noise sources in the system, from the varactor and the tank resistance for example. The single-ended amplitude is ~710 mV as compared with our estimate of 860 mV, so we are around 2 dB off there.

Next Steps: The next steps in the development would be to run through all process, voltage, and temperature corners to make sure we meet the specifications everywhere. After that, a full layout including parasitic elements should be done. The size of the capacitor is likely to need some adjustment after the physical design is completed.

Summary: The main take-home point from this design example is to do one's homework before simulating. A proper estimation analysis gives a great starting point for the simulation phase where, most of the time, we just need to fine-tune the sizing.

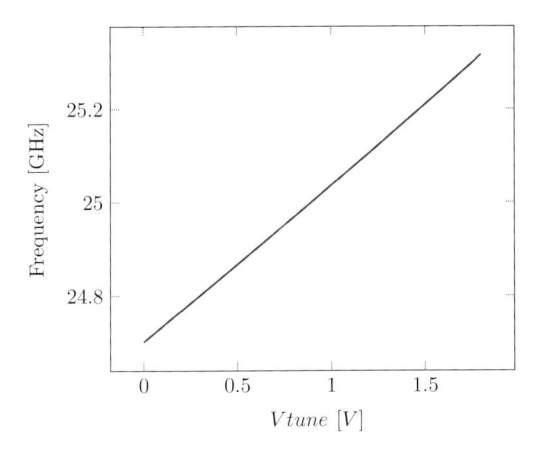

Figure 7.11 Oscillation frequency vs control voltage.

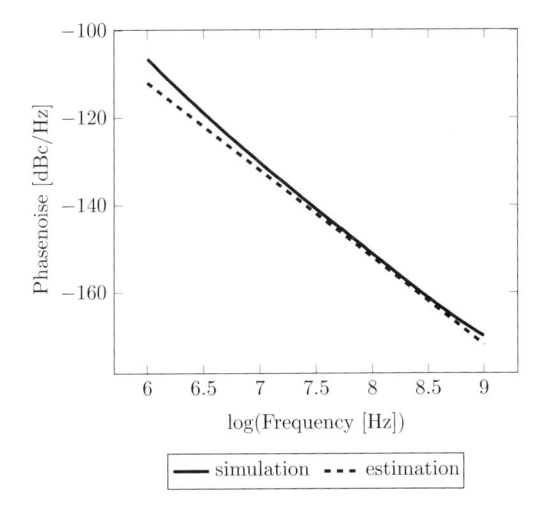

Figure 7.12 Phase noise of oscillator nodes.

However, be aware that VCO design is a rich subject with lots of available topologies and the performance criteria we have focused on here are just the very basics. For a thorough analysis of oscillator design, please see [14]. Having cautioned the reader properly on the limits of the analysis, we have demonstrated here that the path to deeper understanding lies along the route we have just taken.

7.5 Analog-to-Digital Converters

Introduction

Analog-to-digital converters (ADCs) are one of the key building blocks in today's circuitry (see [15–19]). They are part of almost every large integrated circuit in one form or another. The reason is obvious: we live in an analog world, but our circuitry is mostly good at handling digital information, so one needs an interface between them. Over the years many different kinds of architectures have been invented, and we will just briefly mention them here. One common theme in all these systems is sampling, where a signal is looked at briefly and then converted into a digital word and then the process repeats.

This section will first discuss simple models of ADCs as a whole. We start with the most basic of models where there is no explicit sampler. We then continue with a simple model of a sampled ADC followed by a discussion of architectures, performance criteria and a design example.

The general discussion of ADCs is followed by a section on sampling. The effect of sampling can cause unexpected results to the uninitiated, and we will discuss how to simplify the analysis of the sampling process so that a clearer picture will emerge. We will start with voltage sampling, which is perhaps the most common, and continue with charge sampling, a somewhat less used technique. Incidentally, sampling can be viewed as an up–down converter and we will take a quick peek at this effect also in the exercises.

Basic ADC Model

Let us consider a simple model of an ADC and approach it with the estimation analysis we are discussing in this book. We will first make a simple model and see what we can learn from it regarding an ADC's performance. In particular we will calculate quantization noise using this simple model.

Simplify To avoid unnecessary details, let us assume we have a data converter that outputs signed integers in the range $-2^{N-1} + 1 \rightarrow 2^{N-1}$. This will result in 2^N output levels. We further assume

1. The input to this ADC is a sine wave

$$f_{in} = A \sin \omega t,$$

where

$$A = 2^{N-1}$$

2. $N \gg 1$ so that the input signal can be approximated with a straight line in between transitions.

The output of this simple model is now simply

$$f_{out} = \text{int}(f_{in} + 0.5)$$

The 0.5 comes from the assumption that the output is the closest integer level to the input signal. We will now see a step function response at the output as in Figure 7.13.

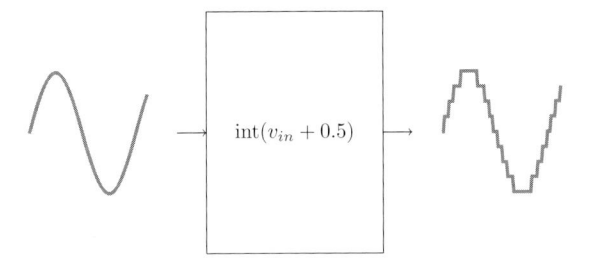

Figure 7.13 Basic ADC functionality.

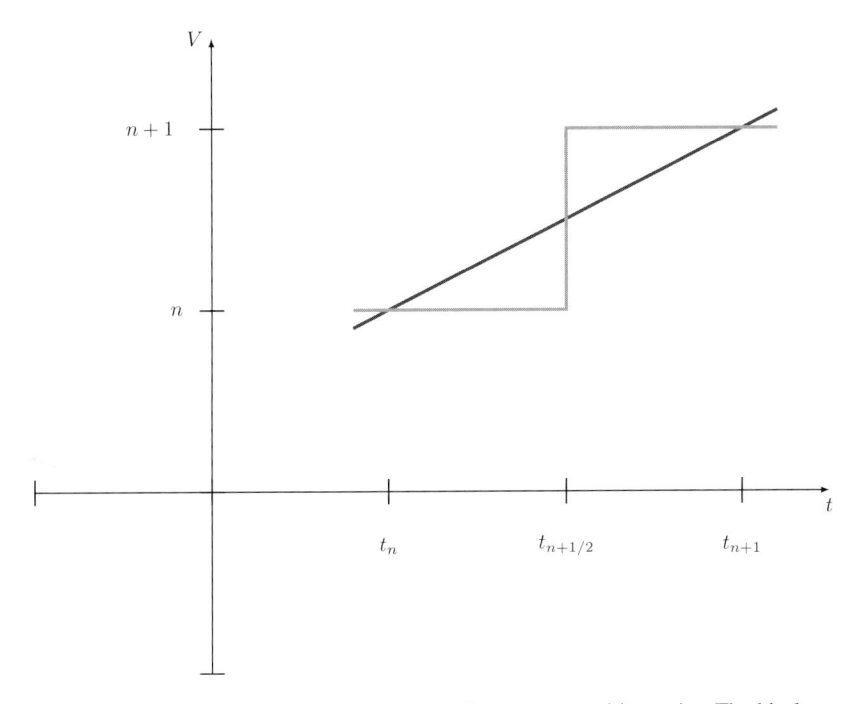

Figure 7.14 Close-up of ADC model output vs input at a transition point. The black curve is the input signal and the gray curve is the output signal.

Clearly the original tone is in there, but there are also other signals present. Let us find those

$$f_{noise} = f_{out} - f_{in} = \text{int}(f_{in} + 0.5) - f_{in}$$

Figure 7.14 shows this graphically.

The noise is simply a sawtooth function over the transition time that starts at 0, goes to -0.5 at the midpoint, $t = t_{n+1/2}$, where the transition occurs, which adds a 1 to the noise which is then reduced back to 0 at $t = t_{n+1}$. Mathematically we have

$$f_{noise} \approx \sum_n \left(-\frac{(t - t_n)}{(t_{n+1} - t_n)} + 1 \cdot \theta(t - t_{n+1/2}) \right)$$

where we have used the Heaviside function θ.

System Aspects

Solve Let us calculate the power in one of these windows

$$P_{noise,n} = \int_{t_n}^{t_{n+1}} f_{noise}^2 dt = 2\int_{t_n}^{t_{n+1/2}} \left(\frac{(t-t_n)}{(t_{n+1}-t_n)}\right)^2 dt = \{t' = t - t_n\}$$

$$= 2\int_0^{\Delta t_n/2} \left(\frac{t'}{\Delta t_n}\right)^2 dt' = 2\left[\frac{t'^3}{3(\Delta t_n)^2}\right]_0^{\Delta t_n/2} = \frac{1}{12}\Delta t_n.$$

Over a period, $\Delta t = T$, we get a total noise

$$P_{noise} = \frac{1}{12}T.$$

The power of the sine wave over the same period is simply

$$P_{signal} = \frac{A^2}{2}T.$$

We have then a signal-to-noise ratio of

$$SNR = \frac{TA^2/2}{T/12} = 6A^2 = 6\,2^{2N-2}.$$

A more familiar form of this is the dB version

$$SNR_{dB} = 10\log\left(6\,2^{2N-2}\right) = 10\log 6 + 10\,\log\left(2^{2N-2}\right)$$
$$= 10\log 6 + 10\,(2N-2)\log(2) \approx 7.781 + 3.01\,(2N-2) = 6.02N + 1.76.$$

Verify With N bits the quantization noise from an ADC is such that the SNR becomes

$$SNR_{dB} = 6.02N + 1.76.$$

This is a well-known formula and can be found in pretty much all textbooks on ADCs. We see again that by making a really simple model of an ADC we can learn something fundamental about their properties.

Evaluate When we discuss noise of an ADC, it is often dominated by the quantization noise. There is little point in making an ADC completely dominated by thermal noise since we can get similar performance simply by decreasing the resolution, which can lead to a significant reduction in design time (and power!).

Key Concept

The quantization noise of an ADC results in a signal-to-noise ratio of

$$SNR_{dB} = 6.02N + 1.76$$

ADC Model with Sampling

The basic model just presented has some obvious limitations, the most severe of which is that the sampling rate is dependent on the signal itself. This means it will be very difficult to find out its frequency content a priori, and the downstream processing will need to resample the output or some such scheme. It is also clear from the very simplified model just presented that in a real world ADC there is a finite time needed for the circuitry to convert the signal to a digital word. If this finite time is an appreciable portion of the signal period, the circuitry can quickly get confused and signal loss/distortion effects and other degradations can occur. The classic remedy for both of these problems is a uniform sampler in time, where the input signal is held for a certain amount of time, giving the circuitry a chance to convert the signal undisturbed. Sampling is almost always done at a fixed frequency and there are a number of ways to accomplish this with circuitry. We will study two such common methods in this chapter from a systems perspective: voltage sampling and charge sampling. The effect of sampling on a signal is of course well studied, but here we will include detailed calculations that describe the effect explicitly.

Nyquist Criterion

One effect that should be mentioned here is the Nyquist criterion, which states that with a given sampling frequency, f_s, the signal needs to be within a bandwidth of $f_s/2$. Curiously, this frequency band could be anywhere in the spectrum in principle. If it is within $f_s/2$, it is referred to as the first Nyquist band. If it is between $f_s/2$ and f_s, it is within the second Nyquist band, and so on. We will derive a precise expression for this effect and its cause in section "Voltage Sampling Theory."

> **Key Concept**
>
> Due to folding effects, an ADC needs to have its input signal band limited to within a bandwidth of $f_s/2$ where f_s is the sampling frequency.

In the rest of the chapter we will assume the signal is sampled with a certain constant sampling period, T_s.

To model a uniform sampling window, one traditionally makes an additional assumption to the first two in section "Basic ADC Model."

3. This signal is "active," meaning two consecutive samples have different output words.

This assumption is needed to decouple the concept of noise power from the signal itself. Imagine a DC signal sitting right at the average trigger point. The output would then have no noise. If it sits right at the boundary, it will have maximum noise, 0.5 least-significant-bit (LSB). The noise is then clearly signal-dependent. In a real system this is unrealistic since real world signals one encounters in practice are quite active. These three criteria are referred to as Bennet's criteria (see [15]).

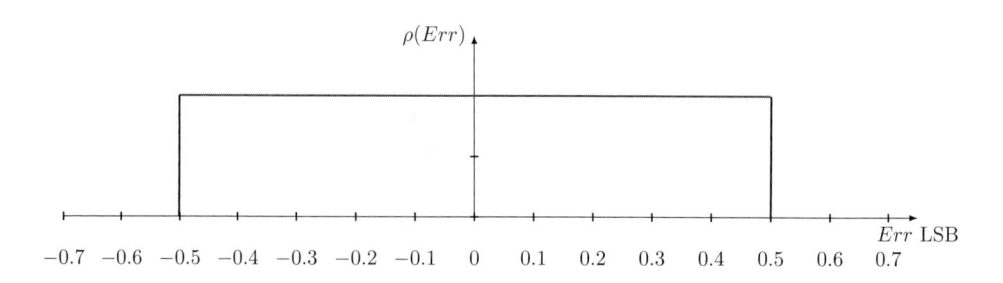

Figure 7.15 Uniform transition probability.

At each sampling point we now make the simplifying assumption that the quantization error is uniformly distributed over the range + 0.5 LSB shown in Figure 7.15 (see [15]).

We find for the noise power:

$$P_{noise} = \int_{-1/2}^{1/2} P(E)E^2\,dE = \left\{P(E) = \frac{1}{1/2-(-1/2)} = 1\right\} = \int_{-1/2}^{1/2} E^2\,dE = \left[\frac{E^3}{3}\right]_{-1/2}^{1/2} = \frac{1}{12}[LSB].$$

This is in effect the same calculation we did before, but framed a little differently.

SNR Improvement from Averaging with the Simple Model

Imagine we take the original system and sample it twice as fast, after which we average two consecutive outputs. The overall sampling rate has not changed but something interesting has happened. Let us investigate with a simple model.

Simplify We will look at two consecutive samples, $S_i, S_{i+1/2}$. Each of these can be modeled as

$$S_i = V_i + n_i$$

where V_i is the sampled signal and n_i is the sampled noise. We will simplify by assuming the noise terms are uncorrelated in consecutive samples so their noise power will add up, and furthermore that on average the noise power is the same for all samples.

Solve We find now after adding the two samples

$$S_{i,tot} = S_i + S_{i+1/2} = V_i + n_i + V_{i+1/2} + n_{i+1/2} = V_i + V_{i+1/2} + \sqrt{2}n_i$$

After averaging we find the signal is

$$\langle V_i \rangle = \frac{1}{2}\left(V_i + V_{i+1/2}\right)$$

and the noise

$$\langle n_i \rangle = \frac{1}{2}\sqrt{2}n_i = \frac{1}{\sqrt{2}}n_i$$

The noise power is half of the original, or 3 dB in logarithmic terms, while the signal has the same power. The SNR has improved by 3 dB!

Verify This result can be found in many ADC textbooks [15–19].

Evaluate In general this is a very powerful method to improve SNR if you can handle the speed; each doubling of the sampling rate followed by averaging improves the SNR by half a bit.

$$SNR_K = SNR_1\, K$$

or in dB

$$SNR_{\text{dB},K} = SNR_{\text{dB},1} + 3\,(K - 1)$$

where K is the oversampling factor.

Key Concept

The SNR of an ADC can be improved by oversampling followed by averaging.

Architectures

There are many ways to build ADCs. The modern literature contains a plethora of varieties and we cannot do them all justice here [15–19]. We will just mention some common ones and in the next few sections look at one or two in some detail.

Flash converter: This is simply an input stage driving $2^N - 1$ comparators in parallel. It is the fastest architecture but it has serious size limitations, with every bit increase causing a doubling in size! It is unusual to find implementation higher than six bits and the power consumption it requires can be substantial.

Pipeline converter: The pipeline converter does the conversion in two steps, where each step has fewer bits and thus the combined resolution is higher.

Sigma–delta converter: This is a very powerful way to improve noise performance by up-converting it out of the signal frequency band. Upon low-pass filtering the resulting noise improvement can be dramatic. Its weakness is that it requires oversampling and can thus not be used in really high-frequency applications.

Time-digital converter: This novel idea uses counting of pulse widths to convert analog data to the digital domain. It employs such fancy circuit techniques as time-amplifiers [16].

Successive approximation register (SAR): This implementation does one-bit comparison per sampling clock. It is not a very fast architecture but is extraordinary low power since it is almost exclusively digital in nature.

Time-interleaved converters: This architecture is a combination of sampling + multiplexer + slower ADCs in the back. For modern high-speed data converters it is the most common architecture employed. The slow backend ADC is usually a SAR.

In this chapter we will study an implementation of a flash ADC and look in more detail at the peculiarities of a time-interleaved system.

Performance Criteria

There are many ways to characterize a data converter (see, for example, [15–19]). The application determines which ones should be used. Here we will mention just a few common ones, and for the rest of the chapter we will use the signal-to-noise ratio as the performance criterion.

DC Specification

Resolution: This is the number of output bit lanes for ADCs. It does not necessarily relate to the accuracy of the converter.

Integral nonlinearity (INL): This is the deviation of the output code from a straight line drawn through zero and full scale when the input is a straight DC ramp.

Differential nonlinearity (DNL): This describes the difference between two adjacent code outputs compared with an LSB step size.

Offset: Matching of components is far from ideal in modern integrated circuits. Mismatch effects will cause offset as referred to the input signal. A zero at the input will result in a nonzero equivalent at the output.

Power: The power consumption is often a very critical specification when circuitry is used in battery powered devices.

AC Specifications

The AC or dynamic specifications are often the most highlighted ones, given their reputation as being the hardest to meet. This is often the case, but the same weaknesses can usually be found in the DC specifications.

Signal-to-noise ratio (SNR): This is simply the ratio of the signal power to the power of everything else, excluding harmonic distortion.

Total harmonic distortion (THD): This is the sum of all harmonic powers divided by the power in the main tone, expressed most commonly as a percentage (%). We will use THD in dB.

Signal-to-noise and distortion ratio (SNDR): This is simply the ratio of the signal power to the power of everything else, including harmonic distortion.

Effective number of bits (ENOB): The definition of ENOB is somewhat confusing. There is a formal IEEE definition, but most current literature uses a simpler one defined as

$$ENOB = \frac{20 \log SNDR - 1.76}{6.02}$$

Spurious-free dynamic range: This is the difference in dB between the main tone and the highest spur in the spectrum.

Bit error rate (BER): This the ratio of conversion errors over the number of conversions done: see Chapter 3.

In most modern papers describing ADC performance the discrete Fourier transform of the output signal is used:

$$h(k) = \sum_{n=0}^{N-1} H(t_n) e^{-j2\pi \frac{k}{T} t_n}, t_n = \frac{n}{N} T, k \in \left[-\frac{N}{2}, \frac{N}{2} - 1 \right], N \text{ even}$$

One can use this formula to investigate issues such as SNR, SNDR, THD, and other such issues that occur with reasonable frequency. Things such as $1/f$ noise, and glitches including missing codes, either require excessive sampling times or need to be investigated with other means. Using these kinds of transforms can be very revealing of the system's performance. Be on the lookout for odd noise floor behavior, and unusual output signal power. These can be indicative of unhealthy circuit behavior. If the capture time is not long enough, the detrimental effects of jitter can be hidden in the main tone. Apart from some of the rare defects mentioned previously, the Fourier transform of the signal is very revealing of the full circuit behavior.

For our purposes we will use SNR, SNDR, and THD as key specifications to illustrate the use of estimation analysis in the design of data converters.

Interleaving ADC

Many modern high-speed ADCs use time-interleaving topologies to achieve high sample rates. An example is shown in Figure 7.16.

The idea is to mux the input sampled data to several slower-speed ADCs that output their data in a synchronous manner. This output is aligned timing-wise to achieve the designed sample rate. Various imperfections in the slower-speed ADCs will affect the output result in predictable way, and we will study a simplified version of such an ADC in this section.

Simplify We make the simplification that there are only two time-interleaved data paths. We will first look at the effect of offset differences between the two digitizers followed by a discussion of gain mismatch. We will also assume there is no thermal noise to worry about.

System Aspects

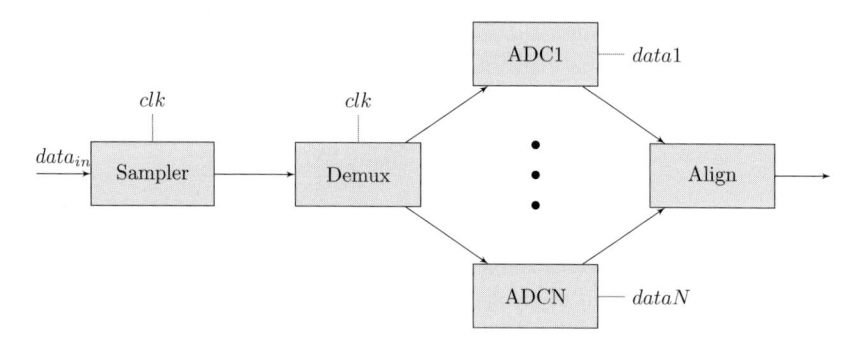

Figure 7.16 Time-interleave topology.

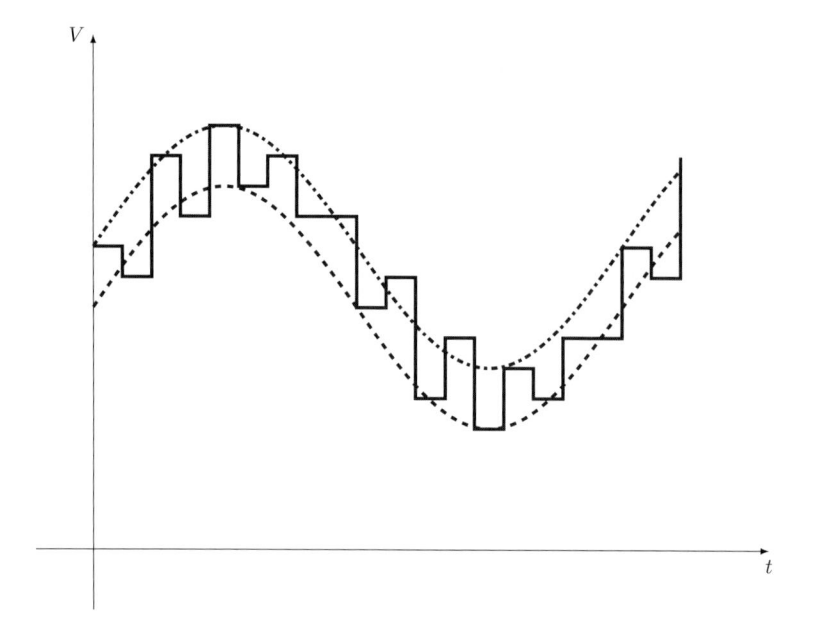

Figure 7.17 Figure showing offset effect. The two dashed curves indicate the offset of the two ADC slices.

$$V_o = \begin{cases} A \sin \omega nT, 2nT \leq t \leq (2n+1)T \\ C + A \sin \omega nT, (2n+1)T \leq t < 2(n+1)T \end{cases}.$$

If we plot this, we find as illustrated in Figure 7.17.

Solve We can now Fourier transform this effect:

$$V_o(f) = \int_{-\infty}^{\infty} V_o(t) e^{-j\omega t} dt = \sum_{n, odd} \int_{-\infty}^{\infty} (C + A \sin \omega nT) e^{-j\omega t} dt + \sum_{n, even} \int_{-\infty}^{\infty} A \sin \omega nT e^{-j\omega t} dt$$

$$= \sum_{n} \int_{-\infty}^{\infty} A \sin \omega nT e^{-j\omega t} dt + \sum_{n, odd} \int_{-\infty}^{\infty} C e^{-j\omega t} dt.$$

7.5 Analog-to-Digital Converters

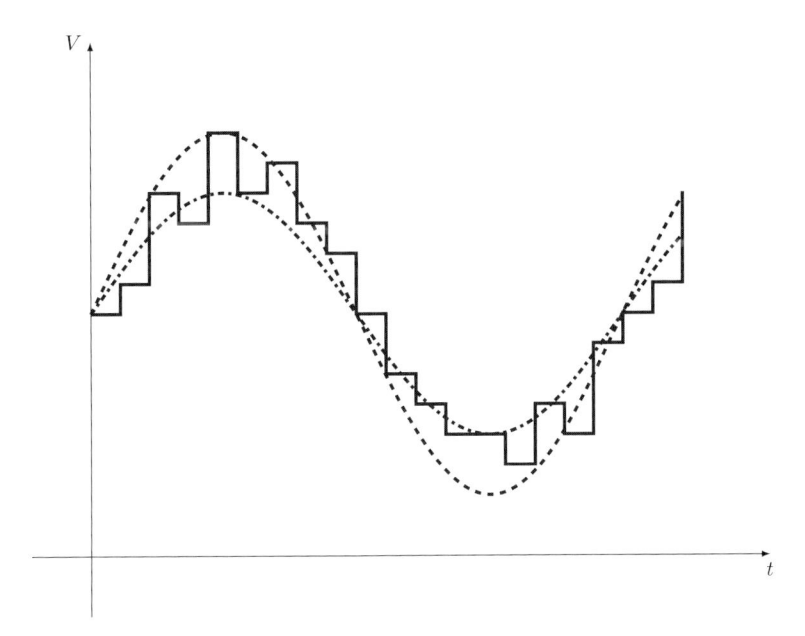

Figure 7.18 Figure showing gain mismatch effect. The two dashed curves indicate the gain of each ADC slice.

In addition to a sinusoid sampled with a period T, there is a DC tone sampled with a $2 \cdot T$ period that will appear at frequencies, $m/2T$, *half* the sampling frequency. This is simply the up-converted version of the offset in one channel.

The effect of gain mismatch can be analyzed similarly: see Figure 7.18.

We now have

$$V_o(f) = \int_{-\infty}^{\infty} V_o(t) e^{-j\omega t} dt = \sum_{n,\,odd} \int_{-\infty}^{\infty} A' \sin \omega_s nT e^{-j\omega t} dt + \sum_{n,\,even} \int_{-\infty}^{\infty} A \sin \omega_s nT e^{-j\omega t} dt$$

$$+ \sum_{n} \int_{-\infty}^{\infty} A \sin \omega_s nT e^{-j\omega t} dt + \sum_{n,\,odd} \int_{-\infty}^{\infty} (A' - A) \sin \omega_s nT e^{-j\omega t} dt.$$

We have a sinusoid sampled with a period T and a sinusoid, with amplitude given by the gain mismatch, sampled at half the sampling frequency. This second term will give rise to tones at $m/2T \pm f_s$.

Verify These are standard results that can be found, for example, in [19].

Evaluate In general for an n-interleaved system, any offset will show up as spurs around F_s/n, and any gain mismatch will show up at $F_s/n \pm f$.

Summary

We have studied simple ADC models and realized some relevant properties by using estimation analysis. Although most of these results are well known, it is hoped that the reader will be encouraged by the methodology used and will be inspired to explore ADCs and their properties on his/her own. By simply pondering the systems under consideration and trying to capture their essence through simple models, much can be learned on one's own.

Example 7.3 Flash ADC design
Since we have discussed a few basic examples of key ADC components, let us put together a flash ADC. This is arguably the simplest and, as such, the fastest of the traditional topologies and will serve as a good example of how to use the estimation analysis to design circuits. It is not necessarily the most power conservative topology, and for such applications where power is paramount the ADC literature has plenty of examples of efficient architectures. Due to space limitations we will only consider a few of the normal specifications here. A full ADC design requires specifications on effects such as offset and power consumption and others that we will leave to the side here. One can, however, easily incorporate these other effects using estimation analysis techniques.

ADC Specifications
Table 7.6 shows the specifications for the ADC.

ADC Design
We will use here the results from the estimation analysis we have done with several circuits along the way in this book, in particular from Chapters 2 and 3, to:

- Get a starting point of the design work from the estimation analysis
- Use this starting point in the simulator where we optimize the parameters

This is another circuit example of how to use estimation analysis to speed up the design work. A good understanding of the basic behavior is essential to a good-quality end

Table 7.6 Specification table for ADC

Specification	Value	Comment
Sampling frequency	12.5 GHz	
SNR	37 dB	
SNDR	35 dB	1 dB off full scale at input
Resolution	6 bits	
Sampler switch ON resistance	10 ohm	When used

7.5 Analog-to-Digital Converters

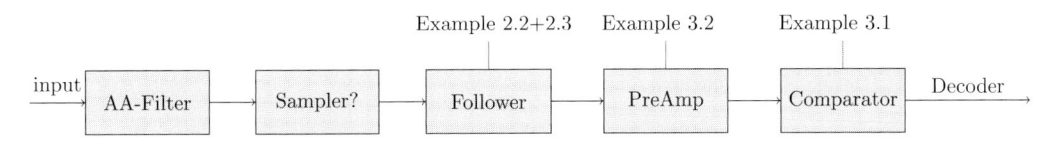

Figure 7.19 Basic ADC topology.

product. The simulator should not give us any surprises; it should just be used to fine-tune the performance, as we will discuss here.

The topology we will use is shown in Figure 7.19. The first block is an anti-aliasing filter that simply ensures the signal coming through to the ADC has the signal limited to the appropriate Nyquist band. This block will not be discussed here at all; instead, we assume the input signal is properly band limited.

We will take the input stage, the follower, from Chapter 2, the buffer and comparator from Chapter 3, and for the sampling switch we will first not use any and then compare the bandwidth to the case where we use an ideal one with a series resistance.

One important issue we have not touched upon when designing the comparator is the kickback phenomenon. The tail reset switch in Figure 3.3 will cause the source voltage of the differential pair to move to ground node vss rapidly. This rapid change in voltage will induce a current through the gate source capacitance of the input transistor and into the driving impedance of the input source.

$$I_{kick} = C\frac{dV}{dt} \rightarrow \Delta V_{in} = I_{kick} Z_{in}$$

This will induce a voltage that will interfere with the precious input signal and can cause significant problems. A simple scaling argument shows

$$\Delta V_{in} \approx 10 \cdot 10^{-15} \frac{0.4}{10 \cdot 10^{-12}} 100 \approx 0.4 \cdot 10^{-1} \quad [V].$$

One is helped to some degree by the differential operation of the circuitry, but if one needs high precision, it may very well be insufficient.

A preferred way to deal with this is to use a preamplifier to isolate the comparator kickback from the input signal. It will also help with offset, but it will cause additional delays and at times it may be unacceptable. One can also use different, higher power, comparators where the circuit is biased the whole time. Here we will use a preamplifier from Example 3.2. We have the topology as outlined in Figure 7.20.

The circuit blocks have already been designed with this system specification in mind, and upon putting everything together and simulating we find the following set of results for different frequencies. *Note, for these simulations nothing was changed compared with the earlier circuitry. Everything was designed/estimated with proper driving source/load impedance and so no adjustments were made!* The resulting spectrums for two separate input tones are shown in Figure 7.21. The final results are shown in Table 7.7.

Table 7.7 Final simulation results for flash ADC

Parameter	Simulation	Estimation	Difference
SNR	37.4 dB	37.9 dB (Quantization noise)	0.5 dB
SNDR	36.8 dB	37.9 dB	1.1 dB @ 6 GHz
BW	7.1 GHz	4 GHz (Known underestimate)	3 GHz

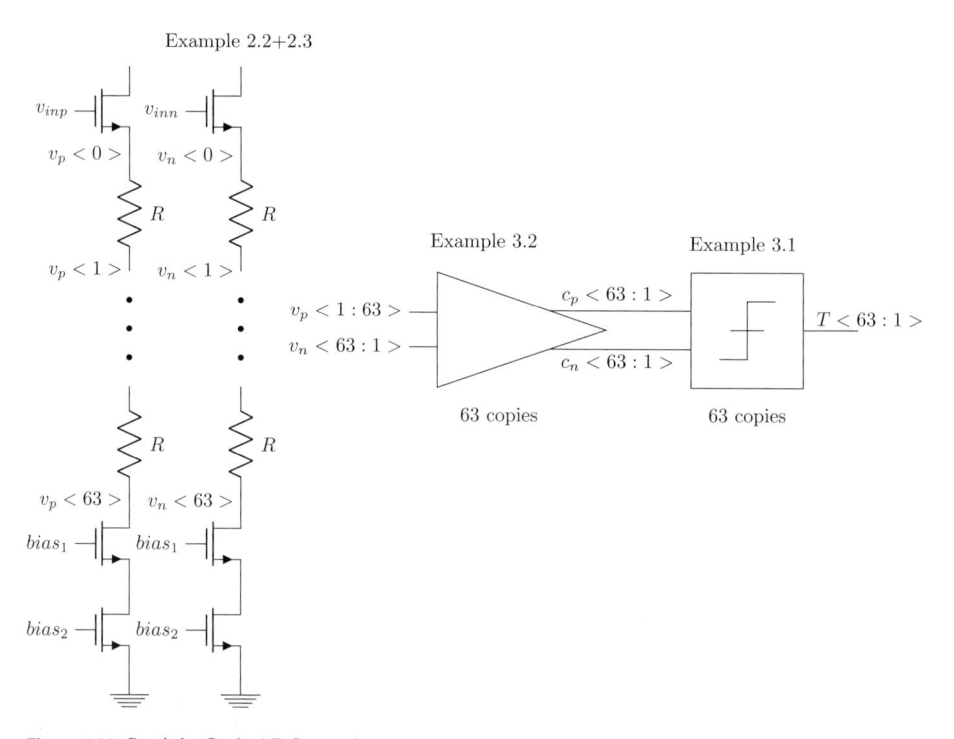

Figure 7.20 Straight flash ADC topology.

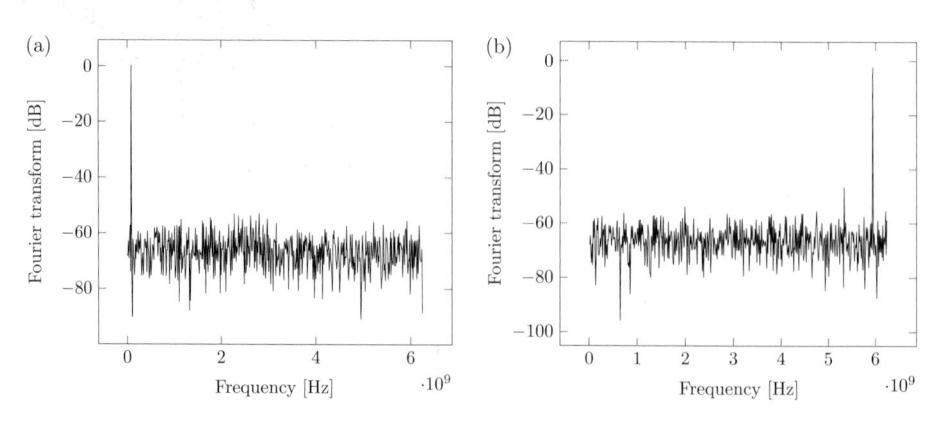

Figure 7.21 Output spectrum of flash ADC at 100 MHz and 6 GHz.

Bandwidth Estimation

We define the bandwidth as the response at the output of the bottom resistor in the string. The bandwidth is set by two properties of the system, first the CD stage output driver and then the fact that we do not use a sampling switch here. These two facts will limit our bandwidth. The aperture window for the comparator is really short, a few ps from our discussion in Chapter 3, section "Comparator Analysis," so it will not impact the bandwidth. The CD stage consists of a string of low-ohmic resistors driving the capacitive input stage of the preamp. There are 63 such RC time constants and the accumulated effect is more significant than just $63 \cdot RC$, since the filters are loading each other. Instead, we can employ estimation analysis to get an idea of the bandwidth.

Simplify Let us approximate the system with a one-pole system, with a resistor equal to the sum of all resistors and a capacitance equal to the sum of all capacitances, where we use a fixed $C = 12$ fF from Example 2.2. The resistor is given by design Example 2.2 plus the follower output resistance $1/g_m = 12$ ohm. This will give us a lower bound on the bandwidth.

Solve The estimated bandwidth is simply

$$f_{BW,lower} = \frac{1}{2\pi RC} = \frac{1}{2\pi(64 \cdot 0.625 + 12) \, 64 \cdot 12 \cdot 10^{-15}} = 4.0 \text{ GHz}.$$

Verify The simulated answer for the RC network itself is

$$f_{BW,sim} = 7.9 \text{ GHz}$$

In the full simulation with active circuitry we find the bandwidth slightly lower due to higher effective capacitance, and we end up with

$$f_{BW\text{-}full,sim} = 7.1 \text{ GHz}.$$

Figure 7.22 shows something interesting. First, Figure 7.22a shows the gain response as we have described it here with a 3 dB point at about 7 GHz. Figure 7.22b shows the same simulation with an ideal sampling switch (10 ohm series resistance) at the gate of the input stage transistor. Here the bandwidth has improved dramatically! The reason is that the resistive ladder now has time to settle since the input is being held for roughly half the sampling period. We will not go into the details of real sampling switch implementations here but hope the reader will be inspired to explore on her/his own.

Distortion Estimation

The distortion will be dominated by the input CD stage. We have seen in Chapter 2 that with a sufficient load, the distortion of a CD stage will be small. In our case we have a large impedance due to the current sink, about 1 kohm, and from the second and third harmonic terms calculated in Chapter 2, Section 2.2 and Example 2.2 we see that the third harmonic term should be

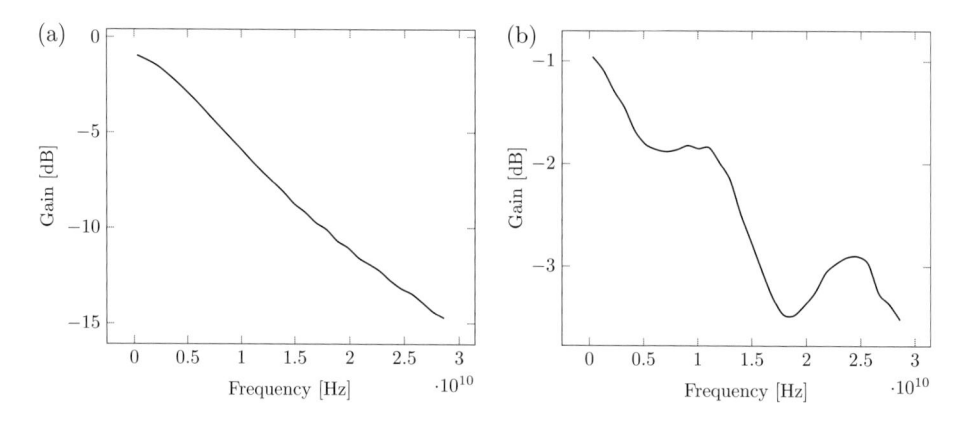

Figure 7.22 Gain transfer function vs input frequency; the frequencies above Nyquist have been measured as the folded down frequencies. The input signal is 1 dB backed off full scale. (a) The gain without an input sampling switch, and (b) the gain response with an ideal input sampling switch in series with a 10 ohm resistor.

$$H_3 \sim \frac{-2(g'_m)^2 + g''_m g_{m,0}}{(g_{m,0})^5 (Z_L)^3} V_{in}^3 \sim 10^{-5}$$

In other words, it is completely negligible. This was also confirmed by simulating low frequencies <100 MHz. For higher frequencies we see from the plots there are some distortion terms showing up, but they are small and will not degrade SNDR by much in this case.

Noise Estimation
The noise is best estimated at the input to the preamplifier. We know from Example 3.1 that the input noise to the comparator is

$$v_{n,rms}^{comp} \sim \sqrt{\frac{I_b}{V_t C_O} \frac{4kT\gamma}{g_{m,1}}} = \sqrt{\frac{3 \cdot 10^{-3}}{0.35 \cdot 25 \cdot 10^{-15}} \frac{4 \cdot 1.38 \cdot 10^{-23} \cdot 300 \cdot 2}{0.024}} \approx 0.7 \text{ mV}$$

The preamplifier's output noise is roughly

$$v_{n,rms}^{pre} \sim \sqrt{(4kTg_{m,p}\gamma R^2 + 4kTR) \frac{1}{2\pi RC_{comp,in}}} \approx 0.9 \text{ mV}.$$

These two noise sources are uncorrelated, and knowing that the preamplifier gain is around 2, we find the total noise at the preamplifier input to be

$$v_{n,rms} = \frac{\sqrt{(v_{n,rms}^{pre})^2 + (v_{n,rms}^{comp})^2}}{2} = 0.6 \text{ mV}.$$

We need to compare this voltage to full scale ADC input, which is 800 mVppd or

$$v_{s,rms} = \frac{0.8}{2\sqrt{2}} = 0.28 \text{ V}$$

Finally, the signal-to-thermal-noise ratio is then expected to be

$$SNR = \frac{0.28}{0.6 \cdot 10^{-3}} \approx 53 \text{ dB}$$

Thermal noise should not be a limiting factor for the performance. We see also from the simulation results that the nonquantization noise is contributing roughly 0.5 dB.

Jitter Impact

We can consider the jitter from the PLL system we designed in Section 7.3. From Figure 7.5 we can estimate a jitter of around 10 fs, but that does not include 1/f contribution at low-frequency offset. Including these sources, the best jitter performance reported in the literature is around 50 fs. Let us assume we can match this number and see what the jitter impact would be on our ADC. At the first Nyquist boundary, we find

$$SNR_{jitter} = \frac{1}{\sigma\omega} = \frac{1}{5 \cdot 10^{-14} \cdot 36 \cdot 10^9} = 0.5 \cdot 10^3 \rightarrow SNR_{dB} = 54 \text{ dB}$$

which is much smaller than the quantization noise. However, at the third and fourth Nyquist bands we may get into trouble if the signal loss is otherwise contained.

Conclusion

We have taken a brief look at the design of a flash ADC. The reader should be aware that a complete ADC design involves far more details. Criteria such as power, offset correction, layout parasitics, error detection, and many others must also be considered. Still, the exercise serves as an example of what one can do with simple estimation analysis. Other important issues can be addressed in the same way. Of particular interest is that the sizing of the various blocks was not changed in simulation; everything had been set up with the correct load and source impedance, and the sizing we had estimated and sometimes verified earlier was correct. So the circuitry worked! This was, of course, a simple example, but imagine the benefits on a larger scale.

Voltage Sampling Theory

Sampling a signal for later digitization is one of the most common operations in any modern IC. How sampling works with up-conversion folding and so on can often be challenging for engineers new to the field. We will use our modeling strategy to shed some light on this phenomenon.

To sample a signal, a precise clock source is used. The quality of the clock source is often described in terms of its jitter, and we will begin this section with a brief discussion of how jitter degrades the signal-to-noise ratio for voltage sampling. We follow this with a discussion of voltage sampling using an ideal switch, where concepts

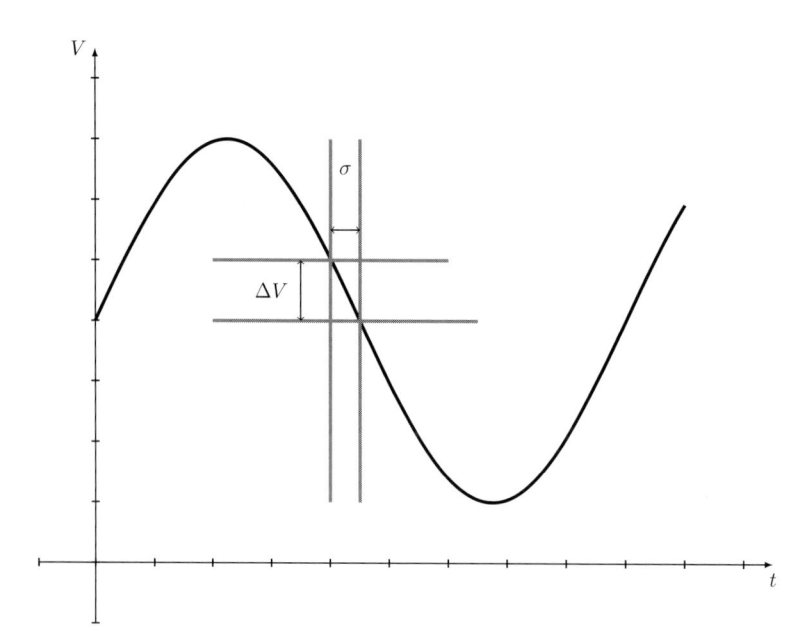

Figure 7.23 Jitter impact on voltage sampling.

such as impulse sampling, track and hold, and so on are defined. This will help us work out some basic mathematical tools we can use for the noisy sampling study in the second section. The simplifications we will use will be stated in the beginning of each subsection and applied to several different subsequent situations. The purpose here is to illustrate how to create simple yet relevant models with some common examples, and in so doing we will present mathematical techniques in some detail.

Jitter

The degradation of signal-to-noise ratio due to jitter in the voltage mode was discussed in section "Phase Noise vs Jitter," where we showed that the equivalent voltage noise follows the well-known jitter "ohms" law (7.3), illustrated in Figure 7.23. We find after some trivial rearrangement that the signal-to-noise ratio of jitter due to voltage sampling is

$$SNR = \frac{A}{\sqrt{\langle v_n(t)^2 \rangle}} = \frac{A}{v_{n,rms}} = \frac{1}{\sigma \omega} \qquad (7.12)$$

Fourier Transforms

In sampling theory we will work with Fourier transforms, and there are a few things worth mentioning before we start. First, we will look at a sampling situation with a fixed sampling period, T. This means we should define the Fourier transform as

$$H(f) = \frac{1}{T} \int_{-\infty}^{\infty} h(t) \, e^{-j\omega t} dt \qquad (7.13)$$

where $\omega = 2\pi f$.

7.5 Analog-to-Digital Converters

Second, we will work with both positive and negative frequencies. For real-valued functions $h(t)$ the following property is easy to prove:

$$H(f) = H^*(-f) \tag{7.14}$$

where * denotes the complex conjugate. For practical calculation we will usually study the positive frequencies and simply keep in mind that the negative ones are the complex conjugate of the positive. It is much easier to work with calculations this way.

Third, instead of using functions such as $\sin \omega t$ one should use Euler's formula for sinusoidal functions:

$$e^{j\omega t} = \cos \omega t + j \sin \omega t$$

From this we see

$$\sin \omega t = \frac{e^{j\omega t} - e^{-j\omega t}}{2j}, \quad \cos \omega t = \frac{e^{j\omega t} + e^{-j\omega t}}{2} \tag{7.15}$$

Working with $e^{j\omega t}$ functions is much easier when calculating Fourier transforms, as should be clear from the definition (7.13).

Working with "negative" frequencies can be a little confusing at first, so let us illustrate how it works with a simple example. Assume we have a function

$$f(t) = A \sin \omega_s t = A \sin \frac{2\pi}{T_s} t$$

where we need to calculate its power. From the basic definition we find the average power is

$$P = \frac{1}{T_s} \int_0^{T_s} f^2(t) dt = \frac{A^2}{T_s} \int_0^{T_s} \sin^2(\omega_s t) dt = \frac{A^2}{T_s} \left[\frac{t}{2} - \frac{\sin 2\omega_s t}{4\omega_s} \right]_0^{T_s} = \frac{A^2}{T_s} \frac{T_s}{2} = \frac{A^2}{2}$$

This is all well known. Now let us take the Fourier transform and look at the spectrum

$$\begin{aligned} F(f) &= \int_{-\infty}^{\infty} f(t) e^{-j\omega t} dt = \int_{-\infty}^{\infty} A \sin(\omega_s t) e^{-j\omega t} dt = A \int_{-\infty}^{\infty} \left(\frac{e^{j\omega_s t} - e^{-j\omega_s t}}{2j} \right) e^{-j\omega t} dt \\ &= \frac{A}{2j} \int_{-\infty}^{\infty} \left(e^{j(\omega_s - \omega)t} - e^{-j(\omega_s + \omega)t} \right) dt = \frac{A}{2j} (\delta(\omega_s - \omega) - \delta(\omega_s + \omega)) \end{aligned} \tag{7.16}$$

where we have used the continuous time definition of the Dirac delta function:

$$\delta(\omega) = \int_{-\infty}^{\infty} e^{-j\omega t} dt$$

We see here we have two frequency tones at $\omega = \pm \omega_s$. Adding their powers, we should end up with the same expression P above:

$$P_{spectrum} = \frac{A^2}{4} + \frac{A^2}{4} = \frac{A^2}{2} = P$$

To summarize:

1. When working with negative frequencies, the tones on the negative frequency side are the complex conjugate of their positive frequency counterparts. We can simply stick to integrating the positive frequency power and multiply the result by two.
2. Alternatively, if we know the power density for positive frequencies, such as resistor noise, we can extend to negative frequencies by simply dividing the power density by 2 and mirror it along the $\omega = 0$ axis.

Armed with these ground rules we can start.

Ideal Sampling Switch

For this section we need to know a couple of Fourier transforms that are helpful, as well as a couple of theorems. We start with something with which most readers will be familiar, impulse sampling, and we continue with track and hold and sample and hold sampling.

Simplify The simple picture we will use throughout the section is pictured in Figure 7.24.

1. We model a sampling switch as an ideal switch in series with a resistor with zero ohm, $R_s = 0$. The loop is completed by a (sampling) capacitor and a driving voltage source.
2. We will also assume for simplicity throughout this section that the on-time is half of the switch period.

Solve With these simplifications we will calculate the Fourier transform for two different situations: impulse sampling and track and hold sampling.

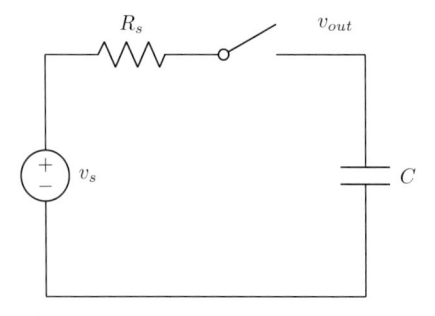

Figure 7.24 Showing ideal sampling switch.

7.5 Analog-to-Digital Converters

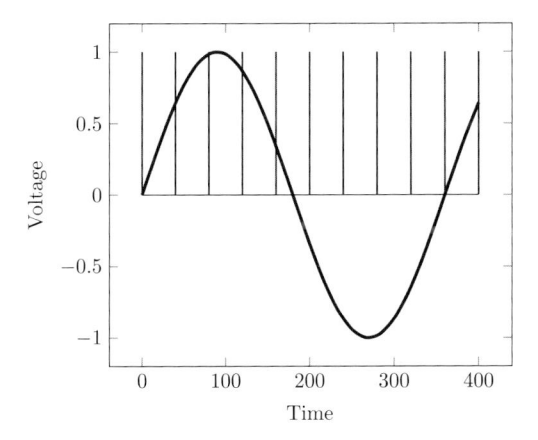

Figure 7.25 Impulse sampling.

Impulse Sampling

Imagine we sample an input sine wave at certain equally spaced points in time only: see Figure 7.25.

Mathematically, this can be expressed as multiplying the input with a series of delta functions:

$$f_{impulse}(t) = A \sin \omega_s t \sum_k \delta(t - kT) = \sum_k A \sin \omega_s t \delta(t - kT)$$

where T is the sampling period. As it stands, it is not very informative. We need to take the Fourier transform to learn more, and in order to do this we first need to take the Fourier transform of the sum of delta function so we can use the convolution theorem (see Appendix B). We have

$$F_\delta(f) = \int_{-\infty}^{\infty} \sum_k \delta(t - kT) e^{-j\omega t} dt = \sum_k e^{-j\omega kT}.$$

The key observation here is that these terms will be zero unless:

$$\omega kT = n2\pi \quad \forall k.$$

The effect of this is a replacement of the sums

$$\sum_{k=-\infty}^{\infty} e^{-j\omega kT} \rightarrow \sum_{n=-\infty}^{\infty} \delta\left(\omega - \frac{n2\pi}{T}\right).$$

We then find

$$F_\delta(f) = \sum_{n=-\infty}^{\infty} \delta\left(\omega - \frac{n2\pi}{T}\right).$$

Now we are ready to use the convolution theorem, and we have for

$$F(f_{impulse}(t)) = F\left(A\sin\omega_s t \sum_k \delta(t-kT)\right) = \int_{-\infty}^{\infty} F(f-f')F_\delta(f')df'$$

$$\int_{-\infty}^{\infty} \frac{A}{2i}(\delta(\omega_s - \omega + \omega') - \delta(\omega_s + \omega - \omega'))\sum_{n=-\infty}^{\infty} \delta\left(\omega' - \frac{n2\pi}{T}\right)df'$$

$$= \sum_{n=-\infty}^{\infty} \frac{A}{2i}\left(\delta\left(\omega - \left(\frac{n2\pi}{T} + \omega_s\right)\right) - \delta\left(\omega - \left(\frac{n2\pi}{T} - \omega_s\right)\right)\right).$$

These delta functions will appear over and over again, so to minimize the clutter we will use the following short forms:

$$\delta_{\omega,n}^{-\omega_s} = \delta\left(\omega - \left(\frac{2\pi n}{T} - \omega_s\right)\right)$$

$$\delta_{\omega,n}^{\omega_s} = \delta\left(\omega - \left(\frac{2\pi n}{T} + \omega_s\right)\right)$$

We then have

$$F_{impulse}(\omega) = \frac{A}{i2}\sum_{n=-\infty}^{\infty} \delta_{\omega,n}^{\omega_s} - \delta_{\omega,n}^{-\omega_s}.$$

We have an impulse train with tones of size $A/2$ around all harmonics of the sampling frequency, as in Figure 7.26. We will see the same kind of calculation again and again all through this section; see [15] for similar discussions.

Track and Hold Sampling

Track and hold is a sampling technique where the input signal is visible at the output with some periodicity. The rest of the time the signal is held at a sampled value:

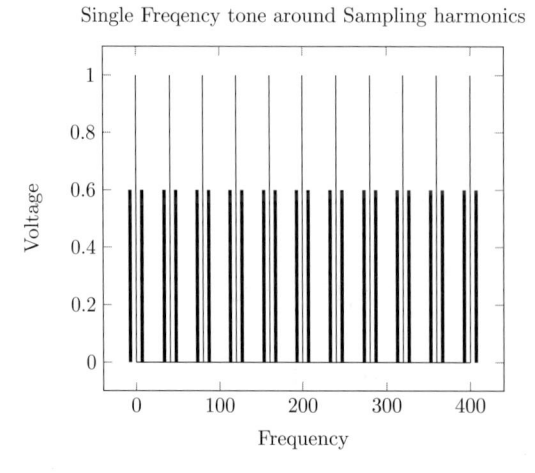

Figure 7.26 Spectrum of impulse sampling.

7.5 Analog-to-Digital Converters

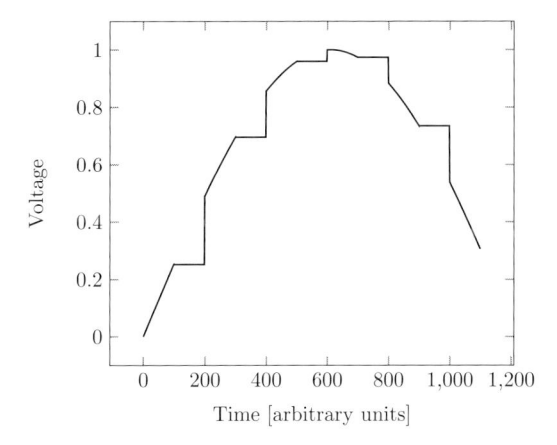

Figure 7.27 A track and hold picture.

Figure 7.27 shows an example. We will first discuss the Fourier transform of the track phase. The hold phase is simpler technically and we leave it as an exercise to the reader. The sum of the track and hold phases gives finally the track and hold result. As a bonus we will also present the sample and hold result.

Fourier Transform of the Track Phase

The track phase can be viewed mathematically as a multiplication of signal by a square wave with amplitude $V = 1$. Such a square wave has the Fourier transform:

$$F_{square}(\omega) = \sum_{k=-\infty}^{\infty} F_k \delta\left(\omega - \frac{2\pi k}{T}\right), \quad F_k = e^{-j\pi k/2} \frac{1}{2} \frac{\sin \pi k/2}{\pi k/2}$$

The Fourier transform of the track phase

$$f_{track}(t) = f_{sig}(t) f_{square}(t)$$

is now:

$$F_{track}(\omega) = \int_{-\infty}^{\infty} F(f - f') F_{square}(f') df'$$

$$= \int_{-\infty}^{\infty} \frac{A}{2j} (\delta(\omega_s - \omega + \omega') - \delta(\omega_s + \omega - \omega')) \sum_{n=-\infty}^{\infty} F_n \delta\left(\omega' - \frac{n 2\pi}{T}\right) df'$$

$$= \sum_{n=-\infty}^{\infty} e^{-j\pi n/2} \frac{1}{2} \frac{\sin \pi n/2}{\pi n/2} \frac{A}{2j} \left(\delta_{\omega,n}^{+\omega_s} - \delta_{\omega,n}^{-\omega_s}\right).$$

We see here that even integers cancel out (except $n = 0$). We end up with

$$F_{track}(\omega) = j\frac{A}{4} \delta(\omega_s - \omega) - j\frac{A}{4} \delta(\omega_s + \omega) + \frac{A}{2} \sum_{n=odd}^{\infty} \frac{1}{\pi n} \left(\delta_{\omega,n}^{+\omega_s} - \delta_{\omega,n}^{-\omega_s}\right).$$

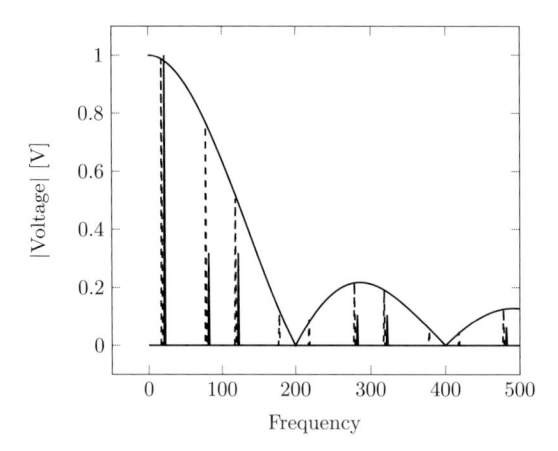

Figure 7.28 Frequency domain representation of track and hold signal. The track phase is indicated in thick black and the hold phase is given by the dashed line. The track and hold responses are offset slightly from each other to show the response more clearly. The overall sinc function is drawn in thin black.

Fourier Transform of a Track and Hold Signal
We leave the hold phase to the reader as an exercise, and when putting all this together in one formula we get

$$F_{T/H}(\omega) = F_{track}(\omega) + F_{hold}(\omega) = \frac{A}{2}je^{-j\omega T/4}\frac{\sin T\omega/4}{T\omega/4}\sum_{m=-\infty}^{\infty}\left(\delta_{\omega,m}^{\omega_s} - \delta_{\omega,m}^{-\omega_s}\right)$$
$$+ j\frac{A}{4}\delta(\omega_s - \omega) - j\frac{A}{4}\delta(\omega_s + \omega) + \frac{A}{2}\sum_{n=odd}^{\infty}\frac{1}{\pi n}\left(\delta_{\omega,n}^{+\omega_s} - +\delta_{\omega,m}^{+\omega_s}\right),$$

and the result can be seen in Figure 7.28.

Sample and Hold Sampling
Sample and hold sampling is similar to track and hold sampling, but the signal is not visible at the output directly; instead, it is held constant after sampling, and depending on the situation it returns to zero after some time or it is held all the way up to the next sample. In this section we calculate the case where the signal is held until the next sample. We leave the more general case to the reader as an exercise.

Fourier Transform of a Sample and Hold Signal
A sample and hold signal where the signal is held constant until the next sample is simply $F_{hold}(\omega)$ with the hold period equal to the sampling period. We can simply do the substitution $T \to 2T$ for $F_{hold}(\omega)$ in exercise 5 and we end up with

$$F_{S/H}(\omega) = \frac{A}{2}je^{-j\omega T/2}\frac{\sin \omega T/2}{\omega T/2}\sum_{m=-\infty}^{\infty}\left(\delta_{\omega,m}^{+\omega_s} - \delta_{\omega,m}^{-\omega_s}\right).$$

A picture of the spectrum in shown in Figure 7.29.

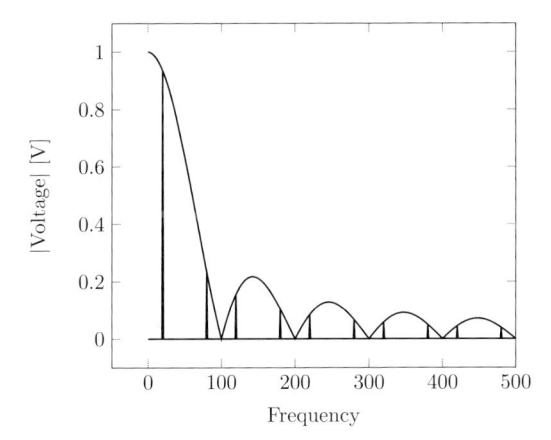

Figure 7.29 Frequency domain representation of sample and hold signal. The overall sinc function is drawn in thin black.

Verify We have studied here some of the basics of sampling and sampled signals. We have seen how, with some basic Fourier series analysis, we can demonstrate a wide range of phenomena. The results presented can be found, for example, in [15].

Evaluate The sampling effect can be simply estimated via Fourier series.

Noise Sampling

We have been looking at the basic sampling effects from a simple perspective, that of a single tone being subject to ideal sampling. What happens if you sample a circuit with noise? We will consider a refined sampling model where there is a noisy resistor in parallel with a capacitor as in Figure 7.24 with the switch permanently closed.

Lossy Resistor in Parallel with a Capacitor
Let us consider the following situation: a resistor in parallel with a capacitor to ground. The voltage source in Figure 7.24 is $v_s = 0$, so the end of the resistor goes to ground and we will study the noise transfer, due to the lossy resistor, of this system to the capacitor when we sample the output node. It is perhaps as simple as one can make it. First, let us study this system without sampling. We can consider a noise voltage sitting in series with the resistor. We find at the capacitor

$$i(t) = \frac{v_n(t) - v(t)}{R} = C\frac{dv}{dt}.$$

Since the noise is better known as a spectrum, we will look at the above in frequency space.

$$\frac{\tilde{v}_n - \tilde{v}}{R} = Cj\omega\tilde{v} \quad \rightarrow \quad \tilde{v} = \frac{\tilde{v}_n}{jRC\omega}\frac{1}{\left(1 + \frac{1}{jRC\omega}\right)} = \frac{\tilde{v}_n}{(jRC\omega + 1)}.$$

The resistor noise is white, and we can now integrate the noise power over frequency to find

$$\int_{-\infty}^{\infty} \left|\frac{\tilde{v}_n}{(jRC\omega+1)}\right|^2 df = |\tilde{v}_n|^2 \int_{-\infty}^{\infty} \frac{1}{\left((RC\omega)^2+1\right)} df = \frac{|\tilde{v}_n|^2}{2\pi RC} \int_{-\infty}^{\infty} \frac{1}{(x^2+1)} dx$$

$$= \frac{|\tilde{v}_n|^2}{2\pi RC} \left[\tan^{-1} x\right]_{-\infty}^{\infty} = \frac{|\tilde{v}_n|^2}{2RC} = \frac{2kTR}{2RC} = \frac{kT}{C}$$

where we have used

$$|\tilde{v}_n|^2 = 2kTR$$

$|\tilde{v}_n|^2$ is defined over negative frequencies, so the power is half the normal textbook version. This is all well. What happens if we sample this system?

Let us assume at time t the capacitor has a voltage $v(t)$, the input is

$$v_{in}(t) = \frac{A_s}{2j}\left(e^{j\omega_s t} - e^{-j\omega_s t}\right),$$

and the voltage on the capacitor is governed by a differential equation derived from

$$Q = CU \quad \rightarrow \quad \frac{dQ}{dt} = I = C\frac{dU}{dt}.$$

We find

$$I(t) = \frac{v_{in}(t) - v(t)}{R} = C\frac{dv(t)}{dt}$$

A simple rewrite shows

$$\frac{dv(t)}{dt} = \frac{v_{in}(t) - v(t)}{RC} = \frac{A_s/2j(e^{j\omega_s t} - e^{-j\omega_s t}) - v(t)}{RC}. \quad (7.17)$$

This is a first-order differential equation that has the solution

$$v(t) = B e^{-\frac{t}{RC}} - \frac{A_s/2 e^{-j\omega_s t}}{RC\omega_s + j} + \frac{A_s/2 e^{j\omega_s t}}{-RC\omega_s + j} = B e^{-\frac{t}{RC}} + D_- e^{-j\omega_s t} + D_+ e^{j\omega_s t} \quad (7.18)$$

$$D_- = -\frac{A_s/2}{RC\omega_s + j} \quad D_+ = \frac{A_s/2}{-RC\omega_s + j}.$$

We take the Fourier of $v(t)$ and find, with $B = 0$ (we assume the initial transient is negligible),

$$F_v(f) = \int_{-\infty}^{\infty} v(t) e^{-j\omega t} dt = \int_{-\infty}^{\infty} \left(D_- e^{-j\omega_s t} + D_+ e^{j\omega_s t}\right) e^{-j\omega t} dt$$

$$= \int_{-\infty}^{\infty} \left(D_- e^{-j(\omega_s + \omega)t} + D_+ e^{j(\omega_s - \omega)t}\right) dt = D_- \delta(\omega_s + \omega) + D_+ \delta(\omega_s - \omega).$$

Ideal Impulse Sampling

Let us sample this with an ideal impulse sampler, where we again use the convolution theorem.

$$F(v(t)v_s(t)) = F(f) = F_v(f) * F_\delta(f) = \int F_v(f-f')F_\delta(f')df'$$

$$\int \frac{A}{2}\left(-\frac{\delta(\omega-\omega'-\omega_s)}{RC(\omega-\omega')-j} - \frac{\delta(\omega-\omega'+\omega_s)}{RC(\omega-\omega')+j}\right)\sum \delta\left(\omega'-k\frac{2\pi}{T_s}\right)df'$$

$$= \sum_{k=-\infty}^{\infty} \frac{A}{2}\left(\frac{\delta^{+\omega_s}_{\omega,k}}{RC(\omega-k2\pi/T_s)-j} - \frac{\delta^{-\omega_s}_{\omega,k}}{RC(\omega-k2\pi/T_s)+j}\right).$$

This simply means the spectrum with a specific sine wave source is shifted around each sampling frequency harmonic. Imagine now that we integrate ω_s over all spectrum, in other words we have a white noise source with noise density $A = 2kTR$. We find

$$V(\omega) = kTR \sum_{k=-\infty}^{\infty} \frac{1}{(\omega+k2\pi/T_s)RC-j} - \frac{1}{(\omega+k2\pi/T_s)RC+j}$$

$$V(\omega) = kTR \sum_{k=-\infty}^{\infty} \frac{2j}{(\omega+k2\pi/T_s)^2R^2C^2+1}.$$

Figure 7.30 shows that by looking at $\omega < \pi/T$, the first Nyquist zone, we see that noise contributions from higher harmonics of the sampling clock extend into the first Nyquist zone. We see that even though we restrict ω to be within the first Nyquist zone, contributions from all harmonics extend into the first Nyquist zone, leaving the total integrated power in the first Nyquist zone equal to the total noise integrated over all frequencies without sampling. In fact, the same can be said about any Nyquist

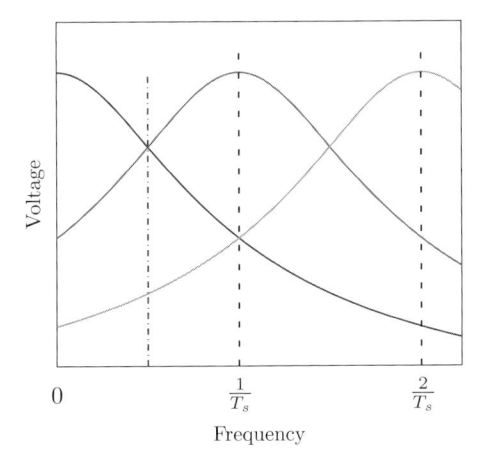

Figure 7.30 First three terms of $|V(\omega)|$. Within the first Nyquist zone the noise from the $m = +1,2$ terms are added to the noise, resulting in an increase in the total noise. The dashed lines indicate harmonics of the sampling frequency and the dash-dotted line the first Nyquist zone.

zone. All Nyquist zones contain all noise power in the ideal impulse sampling case. For a sample and hold situation, it is a little bit different, something we will explore next.

Finite Pulse Width Sampling
Let us sample this with a square wave where the switch is on for a time $T_s/2$ and the rest of the time the output is zero.

$$F(v(t)v_s(t)) = F(f) = v(f) * F_{square}(f) = \int v(f - f') F_{square}(f') df'$$

$$\int \frac{A}{2} \left(-\frac{\delta(\omega - \omega' - \omega_s)}{RC(\omega - \omega') - j} - \frac{\delta(\omega - \omega' + \omega_s)}{RC(\omega - \omega') + j} \right) \sum_{k=-\infty}^{\infty} F_k \delta\left(\omega' - \frac{2\pi k}{T} \right) df'$$

$$= \sum_{k=-\infty}^{\infty} \frac{A}{2} e^{-i\pi k/2} \frac{1}{2} \frac{\sin \pi k/2}{\pi k/2} \left(\frac{\delta_{\omega,k}^{+\omega_s}}{RC(\omega - k2\pi/T_s) - j} - \frac{\delta_{\omega,k}^{-\omega_s}}{RC(\omega - k2\pi/T_s) + j} \right)$$

After integrating the noise source again, we find the following spectrum

$$V(\omega) = kTR \sum_{k=-\infty}^{\infty} e^{-i\pi k/2} \frac{1}{2} \frac{\sin \pi k/2}{\pi k/2} \frac{2j}{(\omega + k2\pi/T_s)^2 R^2 C^2 + 1}$$

This expression is very similar to what we had before, with the exception of the absence of every even sampling harmonic, except DC.

Verify This is a much-discussed effect, see for example [20].

Evaluate With ideal impulse sampling, all Nyquist zones contain all noise power in the unsampled system. Finite sampling width results in a notch filter effect excluding some noise around some harmonics of the sampling frequency.

Key Concept

With ideal impulse sampling, all Nyquist zones contain all noise power in the unsampled system. Finite sampling width results in a notch filter effect excluding some noise around some harmonics of the sampling frequency.

Charge Sampling Theory

Charge sampling is a less frequently used technique to sample a signal. It was first proposed a few decades ago and although the utilization of this technique is not that extensive, it has some advantages that are worth discussing. We will follow [21] in the beginning and extend that discussion to more general situations. Since this technique is used less often, we will spend quite a bit more time on the mathematical derivations. We

7.5 Analog-to-Digital Converters

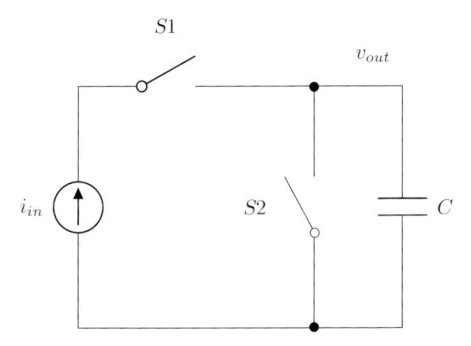

Figure 7.31 Ideal current switch. © [2005] IEEE. Adapted, with permission, from IEEE Trans Circ. Systems.

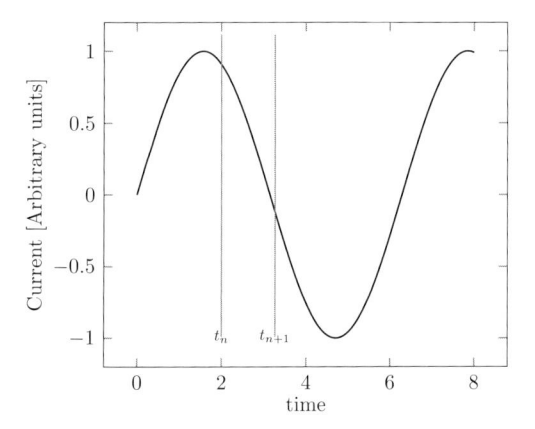

Figure 7.32 Ideal charge sampling in time domain.

will use estimation analysis and start with some simplifications, followed by the now familiar Solve, Verify, and Evaluate discussions.

Simplify Let us consider an ideal charge sampling model starting with the discussion in [21] and Figure 7.31.

We will further simplify the discussion by assuming $S1$ is always on and $S2$ is an ideal switch that clears the capacitance in no time, with a period T_s. When looking at a sine wave, we then have current integration from a time, t_n to a time $t_{n+1} = t_n + T_s$: see Figure 7.32.

Solve Using

$$i_{in}(t) = I_0 \sin \omega t \tag{7.19}$$

we find the sampled charge

$$C_n = \frac{A}{T_s} \int_{t_n}^{t_{n+1}} \sin \omega t\, dt = \frac{-A}{T_s \omega} \left[\cos \omega t \right]_{t_n}^{t_{n+1}} = \frac{A}{T_s \omega} \left(\cos \omega t_n - \cos \omega t_{n+1} \right)$$

where we have defined, $A = I_0 \cdot T_s$, as the charge amplitude. Using

$$t_n = n\, T_s$$

we get:

$$\begin{aligned}
C_n &= \frac{A}{T_s \omega}(\cos \omega n T_s - \cos \omega(n+1)T_s) \\
&= \frac{A}{T_s \omega}\left[-2 \cdot \sin\left(\frac{1}{2}(\omega n T_s + \omega(n+1)T_s)\right) \sin\left(\frac{1}{2}(\omega n T_s - \omega(n+1)T_s)\right)\right] \\
&= \frac{A}{T_s \omega}\left[2 \cdot \sin\left(\omega n T_s + \frac{1}{2}\omega T_s\right) \sin\left(\frac{1}{2}(\omega T_s)\right)\right] \\
&\approx A \sin \omega n T_s, \quad \omega T_s \ll 1.
\end{aligned}$$

(7.20)

We see here that for low frequencies we recover the sampled signal. For higher frequencies there is a roll-off determined by the last sine factor. At Nyquist, $\omega T_s = \pi$ this roll-off has an amplitude of $A \cdot 2/\omega T_s = A\, 2/\pi$ or -3.9 dB loss compared with DC.

When jitter is added into the picture we get:

$$t_n = n\, T_s + \sigma_n$$

where

$$\sigma_n = \sum a_m \sin(\omega_m t + \theta_m)$$

we get:

$$\begin{aligned}
C_n &= \frac{A}{T_s \omega}(\cos \omega(n T_s + \sigma_n) - \cos \omega((n+1)T_s + \sigma_{n+1})) \\
&= \frac{A}{T_s \omega}[\cos(\omega n T_s)\cos(\omega\, \sigma_n) - \sin(\omega n T_s)\sin(\omega\, \sigma_n) \\
&\quad - (\cos(\omega(n+1)T_s)\cos(\omega\, \sigma_{n+1}) - \sin(\omega(n+1)T_s)\sin(\omega\, \sigma_{n+1}))].
\end{aligned}$$

Assuming, $\omega \sigma_n \ll 1$, we arrive at

$$\begin{aligned}
C_n &\approx \frac{A}{T_s \omega}[\cos(\omega n T_s) - \omega \sigma_n \sin(\omega n T_s) - (\cos(\omega(n+1)T_s) - \omega \sigma_{n+1}\sin(\omega(n+1)T_s))] \\
&= \frac{A}{T_s \omega}[\cos(\omega n T_s) - \cos(\omega(n+1)T_s) - \omega \sigma_n \sin(\omega n T_s) + \omega \sigma_{n+1}\sin(\omega(n+1)T_s)]
\end{aligned}$$

For low frequencies $\omega T_s \ll 1$ we can now make an interesting observation,

$$C_n \approx A \sin \omega n T_s + A \sin\left(\omega n T_s + \frac{\omega T_s}{2}\right)\frac{(\sigma_{n+1} - \sigma_n)}{T_s} + A\omega \cos\left(\omega n T_s + \frac{\omega T_s}{2}\right)\frac{(\sigma_{n+1} + \sigma_n)}{2}$$

$$C_n \approx A \sin \omega n T_s + A \sin\left(\omega n T_s + \frac{\omega T_s}{2}\right)\frac{(\sigma_{n+1} - \sigma_n)}{T_s} + A\omega \cos\left(\omega n T_s + \frac{\omega T_s}{2}\right)\frac{(\sigma_{n+1} + \sigma_n)}{2}$$

(7.21)

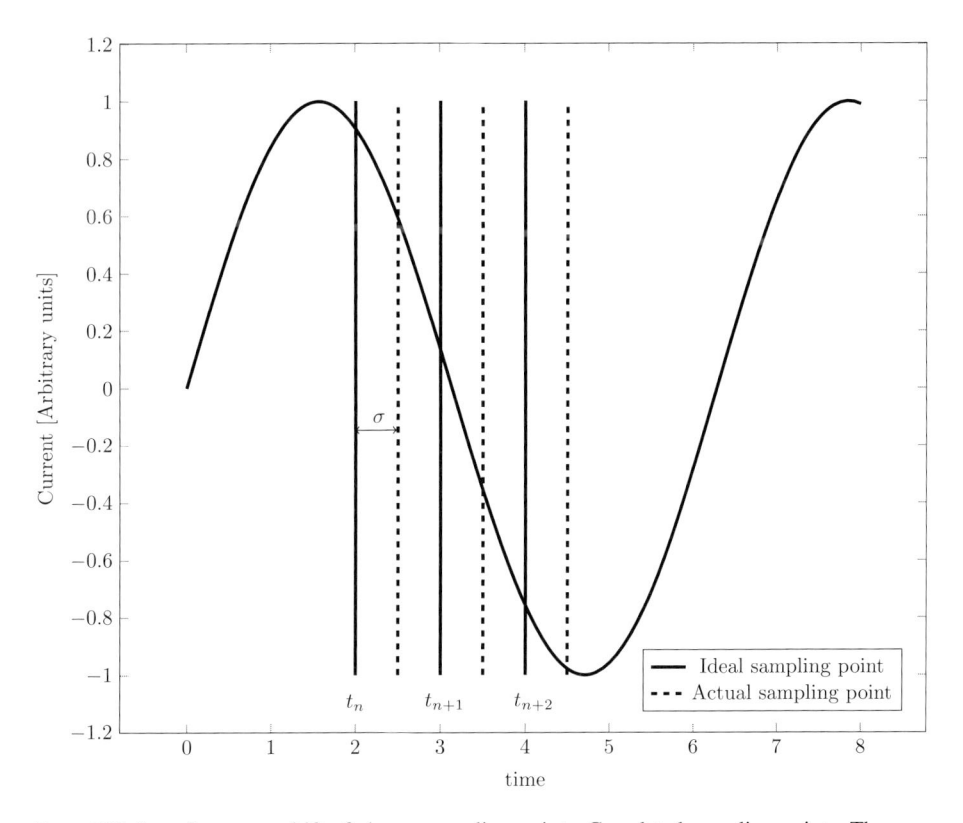

Figure 7.33 Low-frequency drift of charge sampling points. Correlated sampling points. The distance between the sampling points is correct but the average charge sampling time is off by sigma. This situation typically occurs in low-frequency drifts.

The first term is the original signal, while the last two show jitter contributions. Note that if the jitter components are correlated, $\sigma_{n+1} = \sigma_n$ the second term disappears and we have the usual jitter relationship from voltage sampling where the aperture jitter component is proportional to frequency (or the derivative of the signal). One can think of correlated jitter as a *low-frequency drift of the sampling point*: see Figure 7.33. For anti-correlated jitter $\sigma_{n+1} = -\sigma_n$, the third term disappears and we are left with jitter terms that can be called amplitude jitter in that they peak when signal amplitude peaks, shown in Figure 7.34. For uncorrelated jitter, both the second and third terms will contribute, and for small frequencies the second term is the dominant one. We have jitter effects even at DC.

Another way to think of this is in terms of differential signals. Any charge sampling event can be characterized by either the start or end time of the sampling interval *or* the common mode time (average) and the differential time. In other words, we can describe the sampling as a correlated signal plus an anti-correlated signal. These two signals can obviously vary with the sampling time.

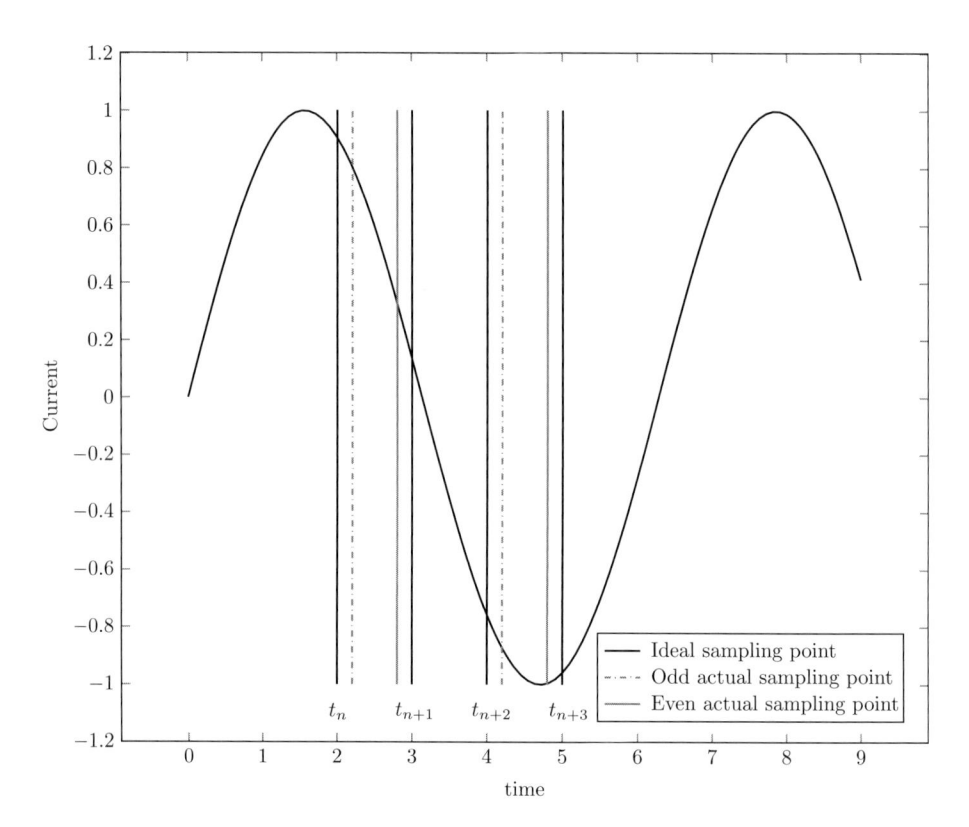

Figure 7.34 Anti-correlated sampling points. The average charge sampling time is correct but the distance between the sampling points is off, causing the slice gain to be off. The odd and even sampling pulse edges move opposite each other.

For general frequencies we can also calculate the response

$$C_n = \frac{A}{T_s \omega}[\cos(\omega n T_s) - \cos(\omega(n+1)T_s) - \omega \sigma_n \sin(\omega n T_s) + \omega \sigma_{n+1} \sin(\omega(n+1)T_s)]$$

$$= \frac{A}{T_s \omega}\left[2*\sin\left(\omega n T_s + \frac{1}{2}\omega T_s\right)\sin\left(\frac{1}{2}(\omega T_s)\right)\right] + J_n \qquad (7.22)$$

where

$$J_n = \frac{A}{T_s}[-\sigma_n \sin(\omega n T_s) + \sigma_{n+1} \sin(\omega(n+1)T_s)]. \qquad (7.23)$$

Verify These calculations are similar to [21].

Evaluate We have identified a couple of different modes of operations, and we will discuss them each in turn. We start with the correlated jitter and follow this by the anti-correlated jitter discussion.

Correlated jitter
When the jitter is correlated we get:

$$J_{n,corr} = \frac{A}{T_s}\sigma_n(-\sin(\omega nT_s) + \sin(\omega(n+1)T_s))$$
$$= \frac{A}{T_s}\sigma_n\, 2\sin\left(\frac{1}{2}\omega T_s\right)\cos\left(\omega n T_s + \frac{1}{2}\omega T_s\right).$$

which is very similar to the main tone except the phase is off by 90 degrees. In other words, it looks like phase jitter. To get the signal-to-noise ratio we need to convert to power and average signal and power over time, T:

$$N_{corr} = \langle J^2_{n,corr}\rangle = \frac{1}{N}\sum_{n=1}^{N}\left(\frac{A}{T_s}\sigma_n\, 2\sin\left(\frac{1}{2}\omega T_s\right)\cos\left(\omega n T_s + \frac{1}{2}\omega T_s\right)\right)^2$$

If we now define an effective jitter function

$$\sigma^2 = \frac{1}{N}\sum_n \sigma_n^2 \cos^2\left(\omega n T_s + \frac{1}{2}\omega T_s\right), \quad N\to\infty \tag{7.25}$$

$$= \left(\frac{A}{T_s}\sigma\, 2\sin\left(\frac{1}{2}\omega T_s\right)\right)^2. \tag{7.26}$$

we see the same roll-off vs frequency as the signal.

For the signal we will also define it similarly

$$S = \langle C_n^2\rangle = \frac{1}{N}\sum_{n=1}^{N}\left(\frac{A}{T_s\omega}\left[2\sin\left(\omega n T_s + \frac{1}{2}\omega T_s\right)\sin\left(\frac{1}{2}(\omega T_s)\right)\right]\right)^2$$
$$= \left(\frac{A}{T_s\omega}2\sin\left(\frac{1}{2}\omega T_s\right)\right)^2. \tag{7.27}$$

We get for

$$\frac{S}{N_{corr}} = \frac{\left(\frac{A}{T_s\omega}2\sin\left(\frac{1}{2}\omega T_s\right)\right)^2}{\left(\frac{A}{T_s}\sigma\, 2\sin\left(\frac{1}{2}\omega T_s\right)\right)^2} = \left(\frac{1}{\sigma\omega}\right)^2. \tag{7.28}$$

This is same expression one would get for pure voltage sampling; see earlier discussion, equation (7.12). Another way to look at this situation is to see that the sampling instance is off by σ_n and if the jitter is slowly changing, the next sample will be off by a similar amount.

> **Key Concept**
>
> Correlated charge sampling jitter degrades SNR the same as voltage sampling.

Anti-correlated jitter
Let us look at the opposite situation where jitter is anti-correlated, $\sigma_n \approx -\sigma_{n+1}$, and the odd and even sample pulses are moving opposite each other.

$$J_{n,\text{anti-corr}} = -\frac{A}{T_s}\sigma_n(\sin(\omega nT_s) + \sin(\omega(n+1)T_s))$$

$$J_{n+1,\text{anti-corr}} = -\frac{A}{T_s}\sigma_{n+1}(\sin(\omega(n+1)T_s) + \sin(\omega(n+2)T_s))$$

$$= \frac{A}{T_s}\sigma_n(\sin(\omega(n+1)T_s) + \sin(\omega(n+2)T_s)).$$

We see in general that we get a sequence of samples with alternating sign

$$J_{n,\text{anti-corr}} = -\frac{A}{T_s}\sigma_n(\sin(\omega nT_s) + \sin(\omega(n+1)T_s))$$

$$= -\frac{A}{T_s}\sigma_n\, 2\,(-1)^n \cos\left(\frac{1}{2}\omega T_s\right)\sin\left(\omega nT_s + \frac{1}{2}\omega T_s\right). \qquad (7.29)$$

For simplicity, let us assume $\sigma_n = -\sigma_{n+1} = \sigma$. We see then that this sequence is generated by multiplying the tone by a square wave with period $2T_s$. The result is a tone at $1/(2*T_s) - \omega/(2\pi)$. In other words, the main tone gets no broadening of its profile. This situation is simply one of timing mismatch between odd/even samples, which will create a tone that looks like an interleave gain error around which the jitter noise moved. The cosine term in the previous expression shows that, close to Nyquist, this term is suppressed.

We can now go through the same calculation that led us to the correlated noise power, N_{corr}. We find

$$N_{\text{anti-corr}} = \left(\frac{A}{T_s}\sigma\, 2\, \cos\left(\frac{1}{2}\omega T_s\right)\right)^2. \qquad (7.30)$$

And

$$\frac{S}{N_{\text{anti-corr}}} = \frac{\left(\frac{A}{T_s\omega}\, 2\, \sin\left(\frac{1}{2}\omega T_s\right)\right)^2}{\left(\frac{A}{T_s}\sigma\, 2\, \cos\left(\frac{1}{2}\omega T_s\right)\right)^2} = \left(\frac{1}{\sigma\omega}\frac{\sin\left(\frac{1}{2}\omega T_s\right)}{\cos\left(\frac{1}{2}\omega T_s\right)}\right)^2. \qquad (7.31)$$

For low frequencies we see the expected dc value, $(T_s/2\sigma)^2$, while for high frequencies we get $\sim 1/(\omega - \omega_s)^2$, which is a remarkable improvement! The improvement in noise is

$$\frac{N_{\text{anti-corr}}}{N_{corr}} = \frac{\left(\frac{A}{T_s}\sigma\, 2\, \cos\left(\frac{1}{2}\omega T_s\right)\right)^2}{\left(\frac{A}{T_s}\sigma\, 2\, \sin\left(\frac{1}{2}\omega T_s\right)\right)^2} = \frac{\cos^2\left(\frac{1}{2}\omega T_s\right)}{\sin^2\left(\frac{1}{2}\omega T_s\right)} = \cot^2\left(\frac{1}{2}\omega T_s\right). \qquad (7.32)$$

We see this ratio is less than 1 above Nyquist/2.

> **Key Concept**
>
> Anti-correlated jitter could provide significant improvement to SNR at frequencies above Nyquist/2.

Uncorrelated jitter

For correlated noise, we get no benefit compared with voltage sampling! For anti-correlated, we get infinite benefit for the tone proper, although the total noise power only rolls down when the main tone is close to Nyquist – still a remarkable improvement in SNR. For uncorrelated jitter, half the jitter will be in the correlated and half will be anti-correlated. We find the total jitter power

$$N_{anti\text{-}corr} + N_{corr} = \left(\frac{A}{T_s}\frac{\sigma}{\sqrt{2}} 2 \cos\left(\tfrac{1}{2}\omega T_s\right)\right)^2 + \left(\frac{A}{T_s}\frac{\sigma}{\sqrt{2}} 2 \sin\left(\tfrac{1}{2}\omega T_s\right)\right)^2$$
$$= \left[\left(\frac{A}{T_s}\frac{\sigma}{\sqrt{2}} 2 \cos\left(\tfrac{1}{2}\omega T_s\right)\right)^2 + \left(\frac{A}{T_s}\frac{\sigma}{\sqrt{2}} 2 \sin\left(\tfrac{1}{2}\omega T_s\right)\right)^2\right] \quad (7.33)$$
$$= \left(\frac{A}{T_s}\frac{\sigma}{\sqrt{2}} 2\right)^2 = 2\left(\frac{A}{T_s}\sigma\right)^2.$$

As the tone frequency changes, the sum of the contributions from the correlated and anti-correlated portions is the same but their relative power changes. For low frequency the anti-correlated jitter is dominant, while for high frequencies the correlated jitter dominates. This noise power can also be found directly from the expression for J_n. We find, noting that both edges contribute equally to noise power,

$$N_{uncorr} = \langle J_{n,uncorr}^2 \rangle = 2 \cdot \frac{1}{N}\sum_{n=1}^{N}\left(\frac{A}{T_s}\sigma_n \sin(\omega n T_s)\right)^2 = 2\cdot\left(\frac{A}{T_s}\sigma\right)^2 \quad (7.34)$$

where we have, similarly to (7.25), defined an effective σ. And we find

$$\frac{S}{N_{uncorr}} = \frac{\left(\frac{A}{T_s\omega} 2 \sin\left(\tfrac{1}{2}\omega T_s\right)\right)^2}{\left(\frac{A}{T_s}\sigma\sqrt{2}\right)^2} = \frac{1}{2}\left(\frac{2}{\sigma\omega}\right)^2 \sin^2\left(\tfrac{1}{2}\omega T_s\right)$$
$$= \frac{1}{2}\left(\frac{2T_s}{\sigma T_s \omega}\right)^2 \sin^2\left(\tfrac{1}{2}\omega T_s\right) = \frac{1}{2}\left(\frac{T_s}{\sigma}\right)^2 \frac{\sin^2(x)}{x^2}, \quad x = \frac{1}{2}\omega T_s \quad 0 \le x \le \frac{\pi}{2}.$$
$$(7.35)$$

The improvement for uncorrelated edge noise compared with correlated noise is:

$$\frac{N_{uncorr}}{N_{corr}} = \frac{\left(\frac{A}{T_s}\sigma\sqrt{2}\right)^2}{\left(\frac{A}{T_s}\sigma 2 \sin\left(\tfrac{1}{2}\omega T_s\right)\right)^2} = \frac{1}{2\sin^2\left(\tfrac{1}{2}\omega T_s\right)} \quad (7.36)$$

which shows that when $(1/2)\omega T_s > \pi/4 \to f > 1/4T_s$ we have an improvement of up to 3 dB at Nyquist. Since we still have jitter at DC, there is a penalty for low frequencies compared with correlated noise.

> **Key Concept**
>
> Uncorrelated jitter can provide up to 3 dB improvement of SNR compared to correlated jitter at Nyquist. However, at DC we will still see an effect of jitter. This is in contrast to voltage sampling.

System Aspects

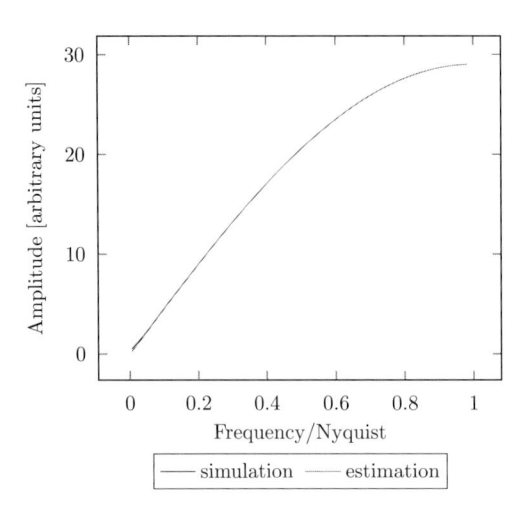

Figure 7.35 Matlab simulation of correlated charge sampling jitter and comparison with the estimated model.

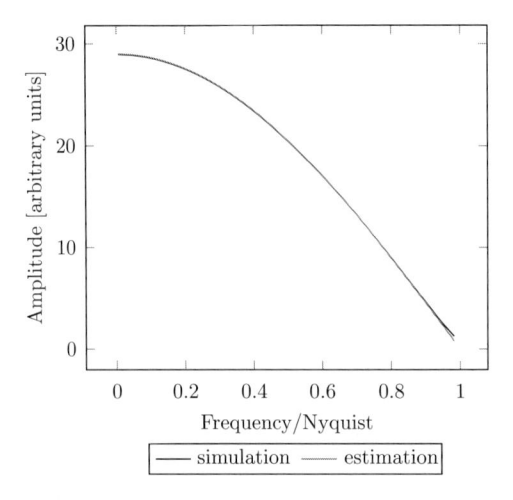

Figure 7.36 Matlab simulation of anti-correlated sampling jitter and comparison with the estimated model.

Verify These two correlation models were simulated with the aid of a Matlab script, and the comparison is shown in Figures 7.35–7.37. We use a simple sinusoidal jitter tone that moves the integration edge up and down in time, similar to the model in section "Phase Noise vs Jitter."

Evaluate The interesting thing to note here is that the anti-correlated signal-to-noise ratio can be significantly higher with the anti-correlated noise at high frequencies. For low frequencies, due to the roll-off of the noise power, it is still better than the correlated noise jitter. This promise caused some excitement early on when it was proposed. However, there is an assumption being made. If we look again at the phase noise response of a typical PLL, in Section 7.1 we found most phase noise was concentrated around the

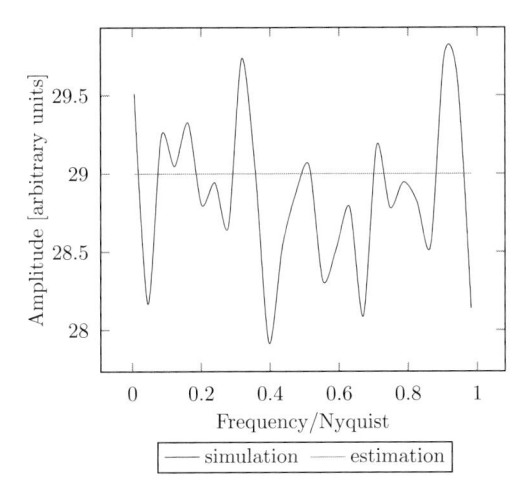

Figure 7.37 Matlab simulation of uncorrelated charge sampling jitter and comparison with the estimated model.

carrier. This causes slow movement of the carrier phase as a function of time. In the context of charge sampling, this means that almost all the jitter is correlated and the SNR degradation due to jitter is more or less identical with voltage sampling. Any portion of the jitter that is uncorrelated will give rise to jitter degradation even for a DC signal. If we can accept the DC degradation, we still need a PLL structure that can spread the noise out so it becomes white or uncorrelated, or we can create a topology that implements an anti-correlated jitter transfer – none of which is obvious, but it is worth some thought.

7.6 Summary

This chapter has explored clock generation in terms of PLLs and its implication in terms of sampling and analog-to-digital converters. We have studied this in the case of jitter in terms of both voltage sampling and charge sampling and have found some interesting relationships, all using simple modeling and estimates followed by verification in terms of literature or simulations. In addition, a few design examples were discussed where the estimation analysis was prominent, and we highlighted how such groundwork can lead to much better understanding of the system and therefore much faster design time.

7.7 Exercises

1. Derive the noise transfer from the VCO, divider, input reference signal, and phase detector block.
2. The open loop response, $O(s)$, for the PLL system in Section 7.3 is defined as

$$O(s) = \frac{F(s)K_{VCO}K_{PD}}{sN}$$

Use this expression to estimate phase margin, defined as the phase of $O(j\omega) + 180$ at the point ω where $|O(j\omega)| = 1$, for the case where

a. $F(s) = \frac{1}{sC}$
b. $F(s) = \frac{1+as}{sC}$

3. You are given a clock reference that has an extraordinary amount of phase noise spread out over the frequency close to the tone. How would you design a PLL that can clean up this noise?
4. You are given a VCO that has very poor phase noise performance. How would you specify the reference clock noise and the PLL bandwidth to fix this?
5. You are given an n bit flash ADC. You are asked to use its components to build an $n + 1$ bit flash ADC. How much would you expect the power consumption to increase? What about area?
6. Calculate the Fourier transform of the hold phase. Verify

$$F_{hold}(\omega) = \frac{A}{2}je^{-j\omega T/4}\frac{\sin T\omega/4}{T\omega/4}\sum_{m=-\infty}^{\infty}\left(\delta_{\omega,m}^{\omega_s} - \delta_{\omega,m}^{-\omega_s}\right).$$

7. How does sampling conversion depend on duty cycle of the sampling clock? We have calculated the case where the hold width is the same as the sampling period in section "Voltage Sampling Theory." Derive the more general case where the signal is held for some portion of the sampling period. The rest of the time output of the switch is zero.
8. Modern simulators have the ability to model nonlinear phenomena where a small signal gets up- or down-converted due to some nonlinear effect. This is often referred to as periodic ac simulation. We can use the model we have built in the voltage sampling section ("Voltage Sampling Theory") to explore this effect analytically, by using the nonideal switch but replacing the noisy resistor with a signal source with a single frequency ω_s and using the track and hold model. First, simplify by only looking at the first harmonic of the sample frequency and investigate how the signal source tone gets up-converted around this harmonic.

7.8 References

[1] M.P. Li, *Jitter, Noise and Signal Integrity at High Speed*, Upper Saddle River, NJ: Prentice Hall, 2007.
[2] R. E. Best, *Phase-Locked Loops*, 6th edn., New York: McGraw-Hill, 2007.
[3] G. Bianchi, *Phase-Locked Loop Synthesizer Simulation*, New York: McGraw-Hill, 2005.
[4] B. Sklar, *Digital Communications*, 2nd edn., Englewood Cliffs, NJ: Prentice Hall, 2017.
[5] J. R. Baker, *CMOS Circuit Design, Layout and Simulation*, 3rd edn., Hoboken, NJ: Wiley-IEEE Press, 2010.
[6] T. Lee, *The Design of CMOS Radio-Frequency Integrated Circuits*, 2nd edn., Cambridge, UK: Cambridge University Press, 2003.
[7] R. Gray, J. Hurst, Lewis, and R. Meyer, *Analysis and Design of Analog Integrated Circuits*, 5th edn., Hoboken, NJ: Wiley, 2009.

7.8 References

[8] W. M. Rogers and C. Plett, *Radio-Frequency Integrated Circuits Design*, New York: Artech House, 2002.

[9] S. Voinigescu, *High-Frequency Integrated Circuits*, Cambridge, UK: Cambridge University Press, 2012.

[10] H. Darabi, *Radio Frequency Integrated Circuits and Systems*, Cambridge, UK: Cambridge University Press, 2015.

[11] B. Razavi, *RF Microelectronics*, 2nd edn., Englewood Cliffs, NJ: Prentice Hall, 2011.

[12] A. Hajimiri and T. Lee, "A General Theory of Phase Noise in Electrical Oscillators," *IEEE JSSC*, Vol. 33, No. 2, p. 179, 1998.

[13] www.maximintegrated.com/en/app-notes/index.mvp/3359.

[14] A. Suarez, *Analysis and Design of Autonomous Microwave Circuits*, Hoboken, NJ: Wiley-IEEE Press, 2009.

[15] J. R. Baker, *CMOS: Mixed-Signal Circuit Design*, 2nd edn., Hoboken, NJ: Wiley-IEEE Press, 2008.

[16] G. Manganaro, *Advanced Data Converters*, Cambridge, UK: Cambridge University Press, 2012.

[17] P. G. A. Jespers, *Integrated Converters*, Oxford, UK: Oxford University Press, 2001.

[18] R. van de Plassche, *CMOS Integrated Analog-to-Digital and Digital-to-Analog Converters*, 2nd edn., Dordrecht, the Netherlands: Kluwer Academic Publishers, 2003.

[19] F. Maloberti, *Data Converters*, Dordrecht, the Netherlands: Springer, 2008.

[20] C. A. Gobet, "Spectral Distribution of a Sampled 1st-Order Lowpass Filtered White Noise," *Electronics Letters*, Vol. 17, pp. 720–721, 1981.

[21] G. Xu, "Performance Analysis of General Charge Sampling," *IEEE Transactions on Circuits and Systems*, Vol. 52, p. 107, 2005.

Appendix A Basic Transistor and Technology Model

This appendix discusses the basic technology parameters we are using in the design examples. It is in itself not an existing technology since that would require various copyright permissions that are not trivial to obtain. Instead the parameters are similar to competitive small geometry bulk CMOS technologies. First we will describe the various dielectric layers in the silicon and then we will through tables and figures list the active device parameters of interest to us here. When the reader applies these techniques to a real process he/she simply needs to fill out the tables and regenerate the transistor functions, perhaps in greater detail than we have space to do here.

A.1   Dielectric Parameters

Table A.1 Dielectric parameters

Parameter	Value	Unit
Silicon substrate thickness	200	µm
Silicon substrate relative permittivity	11.9	
Silicon substrate resistivity	0.1	Ohm m
Epitaxial layer thickness	10	µm
Epitaxial layer relative permittivity	3	
Epitaxial layer resistivity	0	Ohm m
Top-level metal M_{10} material	Aluminum	
M_{10} minimum width	1	µm
M_{10} thickness	2	µm
M_{10} height over substrate	4	µm
M_{10} max current	3	mA/µm
M_9 material	Copper	
M_9 minimum width	0.5	µm
M_9 thickness	0.5	µm
M_9 height over substrate	3	µm
M_9 max current	5	mA/µm
M_8 material	Copper	
M_8 minimum width	0.5	µm
M_8 thickness	0.5	µm
M_8 height over substrate	2	µm
M_8 max current	5	mA/µm
M_2 material	Copper	

Table A.1 (*cont.*)

Parameter	Value	Unit
M_2 minimum width	50	nm
M_2 thickness	100	nm
M_2 height over substrate	300	nm
M_2 max current	1	mA/μm
M_1, M_3–M_7 ignored	N/A	

A.2 Transistor Parameters

The actual simulation uses a full BSIM4 model and the parameters described are derived from such simulations. All transistors in this book have an implicit body to source short that removes the dependence on the back-gate bias on the response also known as the body effect. In practice this means the transistors are laid out in their own well, a so called deep-nwell. This is done for clarity, the body effect can easily be studied with estimation analysis if needed. The basic transistor model we will be using in the hand calculated examples has a drain current that scales like

$$I_d = K \frac{W}{L}(V_G - V_t)^2 \quad (A.1)$$

where K, V_t are assumed given. The source terminal is here assumed to be at ground. Also, the gate capacitance

$$C_{ox} = K_1 WL \quad (A.2)$$

The junction capacitance is assumed to scale with the channel width like

$$C_A = K_2 W \quad (A.3)$$

where we ignore its bias dependence.

From the information in the table we find $g_m r_o \approx 10$ which is fairly typical for this process family.

Figures A.1–A.4 show simple drain current vs voltage simulations of the thin and thick oxide transistors. For small geometry CMOS the PMOS and NMOS response is similar and we assume they are identical for simplicity.

A.3 Noise

Electronic noise shows up in many different forms and can have many different physical origins. For illustrative purposes we only use thermal noise throughout the book. At times other noise sources, and perhaps most significantly, $1/f$ noise needs to be included to have a thorough understanding of a certain situation. We point out to the reader when such noise sources should be considered. For transistors we model noise as an ideal current source across drain and source terminals with a noise density equal to

Appendix A: Basic Transistor and Technology Model

Table A.2 General thin-oxide transistor parameters

Parameter	Value	Unit
K	$0.7 \cdot 10^{-4}$	$\frac{A}{V^2}$
K_1	$30 \cdot 10^{-3}$	$\frac{F}{m^2}$
K_2	$2 \cdot 10^{-10}$	$\frac{F}{m}$
V_t	350	mV

Table A.3 Specific thin-oxide transistor parameters for what is often referred to as a unit transistor in the main text

Parameter, $nf = 10$, $l = 27$ nm, $wf = 1$ μm	Value	Unit
I_d	1	mA
V_g	700	mV
V_s	233	mV
V_d	>250	mV
r_o	1200	ohm
g_m	8	mmho
g'_m	21	mmho/V
g''_m	−14	mmho/V^2
g'_o	−0.025	mmho/V
g'_{om}	2	mmho/V
C_{ox}	8	fF
C_A	2	fF
C_{gd} (fringe)	3	fF
f_t (in saturation $C \approx \frac{2}{3} C_{ox}$)	220	GHz

Table A.4 Specific transistor parameters for 1.5 V device

Parameter, $nf = 10$, $l = 100$ nm, $wf = 1$ μm	1.5 V device	
I_d	1.6	mA
V_g	900	mV
V_s	0	mV
V_d	900	mV
r_o	2k	ohm
g_m	5.8	mmho
g'_m	2.35	mmho/V
g''_m	−5	mmho/V^2
C_{ox}	11.6	fF
C_A	2	fF
C_{gd} (fringe)		fF
f_t (in saturation $C \approx \frac{2}{3} C_{ox}$)	120	GHz

Appendix A: Basic Transistor and Technology Model

Table A.5 Varactor parameters

Parameter unit device	2 V varactor	
Nominal capacitance	10 f	F @V = 0
dC/dV	5 f	F/V
R series	12	ohm

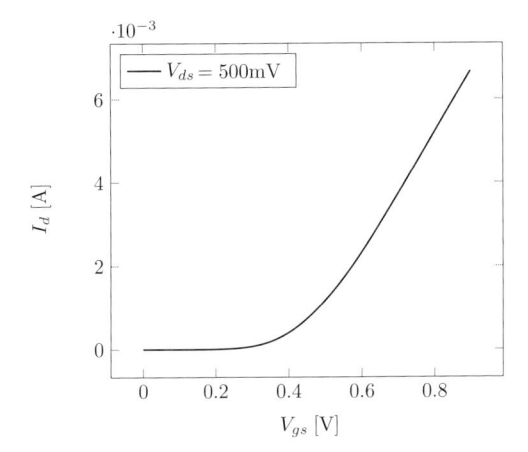

Figure A.1 Drain current vs V_{gs} for transistor with $l = 27$ nm, $w = 1$ μm, $nf = 10$.

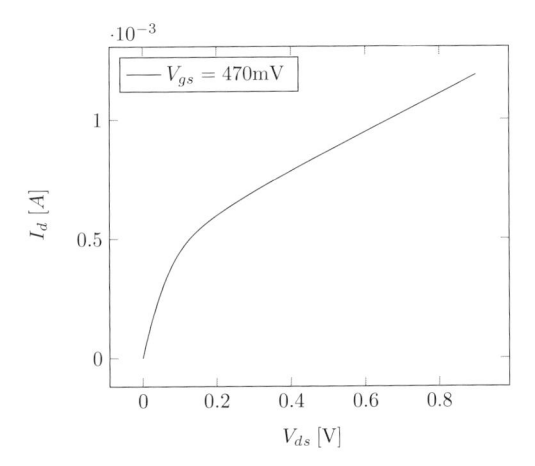

Figure A.2 Drain current vs V_{ds} for same transistor as Figure A.1.

$$i_n^2 = 4kT\gamma g_m \quad [\text{A}^2/\text{Hz}]$$

where k is Avogadros constant, T is temperature in units of Kelvin, [K], γ is a correction factor we assume equal to 2 throughout the book and g_m is the transistors transconductance.

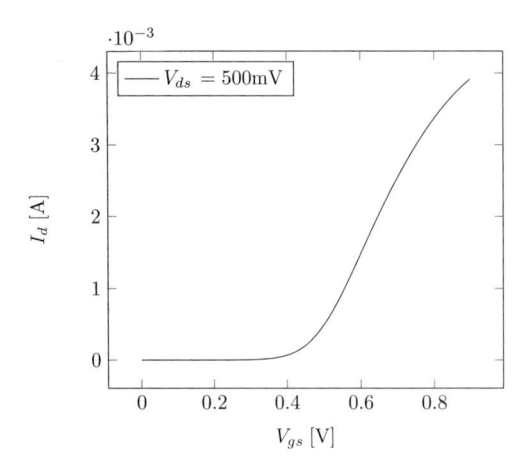

Figure A.3 Drain current vs V_{gs} for thick oxide transistor with l = 180 nm, w = 1 μm, nf = 10.

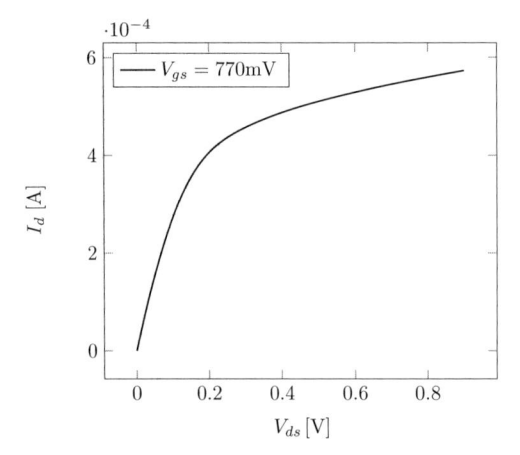

Figure A.4 Drain current vs V_{ds} for same thick oxide transistor.

Appendix B Useful Mathematical Relationships

The formulae in this appendix can be found in many references and online. Here we have used [1, 2] specifically.

B.1 Various Integral Theorems

We assume h, φ, $\boldsymbol{A}$ are smooth scalar or vector functions, V is a three-dimensional volume, S is a closed surface bounding V, with unit outward normal $\boldsymbol{n}$. $\mathcal{F}(h)$ denotes Fourier transform of h.

Parseval's theorem

$$\int_{-\infty}^{\infty} |h(t)|^2 dt = \int_{-\infty}^{\infty} |h(f)|^2 df$$

where $h(f)$ is the Fourier transform of $h(t)$.

Convolution theorem

$$\mathcal{F}(h(t) \cdot g(t)) = \mathcal{F}(h) * \mathcal{F}(g) = \int_{-\infty}^{\infty} h(f - f')g(f')df'$$

Gauss' law

$$\int_V \nabla \cdot \boldsymbol{A}\, dV = \int_S \boldsymbol{A} \cdot \boldsymbol{n}\, da$$

$$\int_V \nabla \varphi\, dV = \int_S \varphi \boldsymbol{n}\, da$$

$$\int_V \nabla \times \boldsymbol{A}\, dV = \int_S \boldsymbol{n} \times \boldsymbol{A}\, da.$$

Below S is an open surface and C the contour bounding it. The normal $\boldsymbol{n}$ to S is defined by the right hand side rule in relation to the sense of the line integral around C.

Stoke's theorem

$$\int_S (\nabla \times A) \cdot n \, da = \oint_C A \cdot dl$$

$$\int_S n \times \nabla \varphi \, da = \oint_C \varphi \, dl$$

B.2 Various Formulas

$$a \cdot (b \times c) = b \cdot (c \times a) = c \cdot (a \times b)$$

$$a \times b \times c = (a \cdot c)b - (a \cdot b)c$$

$$(a \times b) \cdot (c \times d) = (a \cdot c)(b \cdot d) - (a \cdot d)(b \cdot c)$$

$$\nabla \times \nabla \varphi = 0$$

$$\nabla \cdot (\nabla \times a) = 0$$

$$\nabla \times \nabla \times a = \nabla(\nabla \cdot a) - \nabla^2 a$$

$$\nabla \cdot (\varphi a) = a \cdot \nabla \varphi + \varphi \nabla \cdot a$$

$$\nabla \times (\varphi a) = \nabla \varphi \times a + \varphi \nabla \times a$$

$$\nabla (a \cdot b) = (a \cdot \nabla)b + (b \cdot \nabla)a + a \times (\nabla \times b) + b \times (\nabla \times a)$$

$$\nabla \cdot (a \times b) = b \cdot (\nabla \times a) - a \cdot (\nabla \times b)$$

$$\nabla \times (a \times b) = a(\nabla \cdot a) - b(\nabla \cdot a) + (b \cdot \nabla)a - (a \cdot \nabla)b$$

B.3 Laplace Transforms

First let us list the basic relationships and properties of the Laplace transform $\mathcal{L}$:

$$f(s) = \mathcal{L}(F(t)) = \int_{-\infty}^{\infty} F(t) e^{-st} dt$$

$$sf(s) = F'(t)$$

$$\frac{1}{s} f(s) = \mathcal{L} \left(\int_0^t F(x) dx \right)$$

$$f(s-a) = \mathcal{L}\left(e^{at} F(x)\right)$$

Table B.1 Common Laplace transforms

$f(s)$	$F(t)$	Limitations
1	$\delta(t)$	Singularity at +0
$\dfrac{1}{s}$	1	$s > 0$
$\dfrac{n!}{s^{n+1}}$	t^n	$s > 0$, $n > -1$
$\dfrac{1}{s-k}$	e^{kt}	$s > k$
$\dfrac{1}{(s-k)^2}$	te^{kt}	$s > k$
$\dfrac{s}{s^2-k^2}$	$\cosh kt$	$s > k$
$\dfrac{k}{s^2-k^2}$	$\sinh kt$	$s > k$
$\dfrac{s}{s^2-k^2}$	$\cos kt$	$s > 0$
$\dfrac{k}{s^2-k^2}$	$\sin kt$	$s > 0$
$\dfrac{S-a}{(S-a)^2+k^2}$	$e^{at}\cos kt$	$s > a$
$\dfrac{k}{(S-a)^2+k^2}$	$e^{at}\sin kt$	$s > a$
$\dfrac{S^2-k^2}{(S^2-k^2)^2}$	$t\cos kt$	$s > 0$
$\dfrac{2ks}{(S^2-k^2)^2}$	$t\sin kt$	$s > 0$

$$e^{-bs}f(s) = \mathcal{L}(F(t-b))$$

$$f_1(s)f_2(s) = \mathcal{L}\left(\int_0^t F_1(t-z)F_2(z)dz\right)$$

B.4 References

[1] J. D. Jackson, *Classical Electrodynamics*, New York: Wiley and Sons, 1984.
[2] G. Arfken, *Mathematical Methods for Physicists*, New York: Academic Press, 1985.

Index

2-pi-models, 104

active cascode, 34
ADC *see* analog digital converters
ADC model
  basic, 196
  sampling, 199
Ampere's law, 54, 61, 75, 77
amplifier stages
  basic level, 4
  cascaded, 44
  higher level, 31
analog digital converters, x, 195
architectures
  ADC, 201
  PLL, 171

back of the envelope calculations *see* estimation analysis
bandwidth
  estimate, 42
basic PLL equations, 175
basis functions, 162
bipolar, 4
bit error rate, 39, 41
  calculation, 38
  definition, 203
body effect, 4, 235
boundary conditions
  derivation of, 61

capacitance
  calculating, 66
  definition of, 65
  fringe, 28
  one dielectric media, first principle calculation, 68
  simple calculation, 68
  two dielectric media, first principle calculation, 70
capacitor
  circuit element, 67
  serial formula, 72
cascode, 21–22, 28, 31, 33
  active feedback, 33

CD stage, 7, 10
  with resistor capacitor ladder load, 16
CG stage, 5
characteristic impedance, 107–108, 116
charge sampling
  theory, 222
circuit analysis *see* circuit theory
circuit design, 30, 49–50, 85, 90, 101
circuit theory, ix, 51, 109
CMOS inverter
  cross-coupled, 25
  single, 23
CMOS transistor
  inherent gain, 14
  small geometry bandwidth, 10
  small geometry current, 182
common drain *see* CD stage
common gate *see* CG stage
common source *see* CS stage
comparator, 34, 41
  no output load, 40
comparator design, 45
conduction current, 54
conformal coordinate transformation, 95
conformal transformations, 93, 96
continuity equation, 54
convolution theorem, 216, 221
co-planar wave guides, 103
Coulomb gauge, 53
CS stage, 8, 15
current distribution
  perfectly conducting ground plane, 87
  related to inductance, 66
  resistive ground plane, 89
  thin conductors, 93
current mirror, 20
current sheet over ground plane
  first principle calculation, 82
cylindrical symmetry, 2, 58, 74, 76, 85

Delta gap source, 161
design examples, 4, 31, 122, 166, 234
design phase, 3

differential nonlinearity (DNL), 202
differential pair, 18
Dirac delta function
   definition of, 213
distortion, 15, 17, 199, 202, 209
distributed effects, 102

effective number of bits (ENOB), 203
effective radius
   using adjacent circular segments, 126
   using area equivalence, 127
   using maximum length scale, 128
EFIE, 160
electric field integral equation *see* EFIE
electrical length, 108–109, 116, 134
electromagnetism, ix, 50, 101
estimation analysis, 1
estimation calculations, ix
evaluation, 3

field energy, 63
   electrical, 63
   magnetic, 64
fine tune, ix, 3, 191, 194, 207
first principles, ix, 68, 85
five transistor amplifier, 31
Flash ADC, 201, 206
foundry supplied inductor libraries, 104
Fourier transform
   sample and hold signal, 218
   track and hold signal, 218
   track phase, 217
Fourier transform, 212
full scale voltage, 38, 41
full wave approximation
   single frequency, 54

Galerkin's method, 160
gate impedance, 6–7
gauge choice, 53
gauge invariance, 52
gauge potentials, 52–53, 94
gauge symmetry, 53
Gauge theory, 52
gauge transformation, 53
Gauss' law, 62–63
   definition of, 239
Green's function
   1D, 58
ground planes, 102

hand calculation *see* estimation analysis
Helmholz equation, 56–60, 84
high frequency phenomena, 85
high speed integrated circuit
   size of modern, 102
homogeneous equation, 56

ideal impulse sampling, 221
ideal sampling switch, 214
impressed current, 54
impulse sampling, 215
impulse sensitivity function, 187, 189
inductance
   calculating, 73
   definition of, 65
   single straight wire, 74
   single wire over ground plane, first principle
      calculation, 79
   two straight wires, first principle calculation, 74
inductors as circuit elements, 73
initialization phase
   comparator, 35, 42
input impedance, 5–9, 23–25
input pair, 35, 39, 42, 45, 49
input stage
   comparator, 8, 38, 40, 42, 45, 201
   Flash ADC, 207, 209
input transistors *see* input pair
integral nonlinearity (INL), 202
integrated circuits, 4, 53, 56, 70, 101, 119, 122
inter symbol interference (ISI) *see* Jitter:inter symbol
   interference
interleaving adc, 203
inudctor model including
   parasitic capacitand and resistance, 135

jitter, ix, 3, 165–168, 170, 184, 203, 211, 224, 226–231
jitter, 166, 230
   anti-correlated, 227
   correlated, 227
   data-dependent, 166
   deterministic jitter, 166
   duty cycle distortion, 166
   Gaussian, rms, 166
   inter symbol interference, 166
   periodic, 166
   random, 166
   uncorrelated, 229
   uncorrelated bounded, 166
   vs phasenoise, 167

KCL *see* Kirchoff's current law
Kirchoff's current law, 5

Laplace equation, 93
lateral skin effect, 97, 100, 137, 148
LC-VCO, 149, 191
linearization techniques, 4
linearize, 2, 4–5, 21, 175
linewidth, 171
lithography, 102
load resistance, 25, 48

Index

long wavelength
   approximation, 55
   simulators, 152
loop equations, 26
Lorenz gauge, 53

machine learning, 104
magnetic charges, 52
magnetic field integral equation *see* MFIE
Matrix equation, 160
Maxwell's equations, 51, 66–68, 70, 73–74, 85
   1D – solution, 56
   2D – solution, 58
   3D – solution, 59
   external sources, 54
   solutions, 56
metal-insulator-metal capacitor *see* mim capacitor
metastability, 38
method of images, 80
method of moments, 159
MFIE, 160
Miller effect, 9, 27–28
MIM capacitor, 119
MOM capacitor, 121
multi-layer ground planes, 103

narrow-band applications, 8
negative resistor, 7–8, 16, 185
noise analysis, 31, 48
noise optimization, 45
noise transfer, 20, 31–32, 41, 171, 219
nonlinear effects, 4
nonlinear extension, 9, 29
   CD stage, 10
   CS stage, 15
nonideal sampling switch, 219
Nyquist criterion, 199

offset, 202
Ohm's law, 93
one turn inductor, 129
output conductance, 12, 14, 31

Parseval's theorem, 168, 170
   definition of, 239
partial inductance, 122–123
partial inductance of a wire stub pair, 122
PCB
   connection to IC, 102
performance criteria
   adc, 202
   PLL, 172
permeability, 51
permittivity, 51, 55, 70, 72, 85, 102, 136, 139, 234
perturbation analysis, 4
phase locked loop (PLL), ix, 171
phase noise, 187
phenomenological model, 135, 138

pill box, 61
pi-models, 104
pipeline converter, 201
PLL *see* phase locked loop
Poisson equation, 57, 58
positive feedback, 38, 43, 184
potential field, 52, 152
power, 202
printed circuit board (PCB), 102
propagation constant, 106
propagation delay, 23

Q - quality factor
   definition of, 138

rectangular cross section, 126
regeneration phase
   comparator, 37, 42
regularization theory, 104
reset, 35, 40–41, 45, 207
reset phase
   comparator, 35
reset switch, 207
reset switches, 40, 45
RF performance, 104

sampling
   finite pulse width, 222
sampling rate, 41, 199–201
saturation, 5, 16, 23, 27–28, 236
scaling relationships, 42
self-resonance, 134
shielding techniques, 1, 101
sigma-delta converter, 201
signal integrity, 103
signal-to-noise and distortion ratio (SNDR), 202
signal-to-noise ratio
   definition, 202
simple straight wire in free space, 74
simplification, 2
simulation, 12, 15, 46, 81, 125, 131–132, 134, 137, 151–152, 161
simulators, 87, 151
   principles, 151
single transistor gain stages, 4
skin depth, 85
small geometry CMOS, 23, 28, 87, 102, 121
SNR, 41, 166, 198, 200–203, 206, 208, 227–229, 231
   improvement using averaging, 200
S-parameters, 101, 109, 112–114, 116–117
   definition of, 109
spurious-free dynamic range, 203
stability, 176
   feedback systems, 176
Stoke's theorem, 61, 76–77
   definition of, 240
strong-arm comparator *see* comparator

substrate resistance, 136
substrate resistivity, 137, 234
Successive Approximation Register (SAR), 202
surface charge, 63
surface current, 62

Taylor expansions *see* Taylor Series
Taylor series, 10
third harmonic
  differential pair, 19
through-silicon-vias (TSV), 104
time domain, 25, 31, 42, 184, 189, 223
time evolution, 37
time-digital converter, 201
time-interleaved converters, 202
timescale/s, 27, 31, 35–36, 38, 40–43, 45, 48, 234
total harmonic distortion (THD), 202
track and hold, 216, 218
transconductance, 32
transfer function, 10, 15, 17, 25, 29, 171, 175–176, 178, 181–182
transition frequency, 6
transmission lines, 101
  basic theory, 105

two transistor stages, 18
two-turn inductor, 129, 134, 137, 144
two-dimensional solution, 58
two-pole system, 9, 177

VCO
  active design, 193
  frequency of oscillation, 184
  performance criteria, 183
  steady state amplitude, 185
  tank design, 191
vector potential, 52, 56, 60, 66, 76–77, 83, 94, 123, 155, 161
verification, 3
vias, 103
voltage controlled oscillator *see* VCO
voltage field *see* potential field
voltage sampling theory, 211
Volterra series, 9
volume charge, 63
volume current, 61

wavelength, 53, 55, 57, 60, 68–69, 84–85, 88, 93–94, 101–102, 108–110, 117, 151–152, 159